To Jill, Eddy and Andy

Contents

1

Introduction

Welcome to the world of multivariate data analysis. This book introduces the reader to the main areas of the subject without generally going too deeply into the mathematics that underlie the concepts and theories. It should be suitable for university courses in statistics and other disciplines where statistics plays a major role. Hopefully, it will also be useful for others who need an understanding of multivariate analysis techniques in order to analyse data collected during the course of their work.

Exercises are included in each chapter, most being data orientated, but some are mathematically based. Some of the data based exercises can be carried out by hand and a calculator, but others need the use of a computer and a statistical software package. The author believes that, wherever possible, new techniques being learnt should be practised on an example, real or invented, that is small enough to be analysed by hand. Carrying out the calculations manually helps with the understanding and insight into the statistical technique.

Techniques for analysing multivariate data tend to fall into two groups. There are those that are descriptive in nature, helping us understand the data and the interrelationships between the measurements made, and give suitable summaries of the data. The other techniques are model based, giving us a description of the data through a mathematical or probabilistic representation.

1.1 The nature of multivariate data

Data are collected on people, animals, plants or objects. When data are collected for only one variable, then the data are *univariate*. When data are collected for several variables, then the data are *multivariate*. The data collected for a particular person, animal, plant or object, form the *observation* for that particular person, animal, plant or object. Throughout the book, we use the terms *individuals* and *objects* interchangeably as the entities upon which we are collecting data.

Table 1.1 shows a subset of the observations taken during a survey on young people's holidays. The variables of interest are: x_1 – *Gender*, x_2 – *No. in party*, x_3 – *Average amount spent per day (GB pounds)*, x_4 – *Destination*, and x_5 – *Rating within all holidays taken (best = 1, second best = 2, etc.)*.

The variables are of different types. The variable, x_1, 'gender', is *binary*. The variable, x_2, 'number in your party', is *discrete*, taking the values, 1, 2, ... The variable, x_3, 'average amount spent per day', is *continuous*, ranging from zero to some suitable upper limit. The variable, x_4, 'destination', is *categorical* or *nominal*, while the variable, x_5, 'rating within

Table 1.1 Holiday data

Respondent No.	Gender x_1	No. in party x_2	Spend per day x_3	Destination x_4	Rating x_5
1	M	4	50	Spain	2
2	F	2	37	USA	1
3	M	8	42	Turkey	3
4	M	1	25	Argentina	4
5	F	4	65	Spain	2
6	F	2	120	Italy	3
7	M	16	15	UK	1
:	:	:	:	:	:

all holidays taken', is *ordinal*. Multivariate methods need to be able to accommodate these different types of variables.

Chapter 2 discusses vectors and matrices, but we need to make some reference to them now, and it is assumed that the reader has some familiarity with them. The random variables, $x_1, x_2, \ldots, x_5$, each of which is *univariate*, are placed in a column vector, $\mathbf{x}$, called a *vector random variable*, or *random vector*. Each row of data values in Table 1.1 is called an *observation* on the random vector. Generally, the number of univariate variables making up the random vector, $\mathbf{x}$, is denoted by p, and is the *dimension* of $\mathbf{x}$. The total number of observations is generally denoted by n.

The observations on random vector $\mathbf{x}$ are placed in a matrix, $\mathbf{X}$, called the data matrix. Thus

$$\mathbf{X} = \begin{bmatrix} M & 4 & 50 & Spain & 2 \\ F & 2 & 37 & USA & 1 \\ M & 8 & 42 & Turkey & 3 \\ M & 1 & 25 & Argentina & 4 \\ F & 4 & 65 & Spain & 2 \\ F & 2 & 120 & Italy & 3 \\ M & 16 & 15 & UK & 1 \\ \vdots & \vdots & \vdots & \vdots & \vdots \end{bmatrix}.$$

Note that for univariate random variables, we follow the usual convention of using upper case letters to denote 'names' of random variables, but for random vectors, we generally use lower case bold letters. This is to avoid confusion with the upper case bold letters used to denote data matrices.

Often, we will code categorical variables so that the categories are referred to by integers. Thus if gender is coded as *female* – 1, *male* – 0, and destination as *Argentina* – 54, *Italy* – 39, *Spain* – 34, *Turkey* – 90, *UK* – 44, *USA* – 1, then

$$\mathbf{X} = \begin{bmatrix} 0 & 4 & 50 & 34 & 2 \\ 1 & 2 & 37 & 1 & 1 \\ 0 & 8 & 42 & 90 & 3 \\ 0 & 1 & 25 & 54 & 4 \\ 1 & 4 & 65 & 34 & 2 \\ 1 & 2 & 120 & 39 & 3 \\ 0 & 16 & 15 & 44 & 1 \\ \vdots & \vdots & \vdots & \vdots & \vdots \end{bmatrix}.$$

Of course, in any analysis, we must not forget that the integers recorded for categorical variables are not numbers per se, but refer to various categories. However, sometimes we might want to give a score to the categories, such as 0, 1, 2, 3, 4, 5. This is a linear scoring scheme, but others are possible, 0, 1, 4, 9, 16, 25 for instance.

An individual column of the data matrix refers to the data collected on a particular variable, while an individual row corresponds to the data collected on a particular individual or object for all the variables. We will usually refer to columns of the data matrix using the letters i, j, and k, and to rows of the data matrix using the letters r, s, and t. By distinguishing the two sets of letters, it will be clear whether we are referring to individuals (observations) or to variables. The value for the ith variable for the rth observation is x_{ri}, where x_{ri} is the element in the rth row and ith column of $\mathbf{X}$.

The data for the rth individual will sometimes be denoted by the vector, $\mathbf{x}_r$, and is simply the rth row of $\mathbf{X}$. So, for example, $\mathbf{x}_1 = (0, 4, 50, 34, 2)'$ for our data. The data for the ith variable will be denoted by $\mathbf{y}_i$ (or by $\mathbf{x}_i$ if there is no confusion with the ith row), and is simply the ith column of $\mathbf{X}$. Thus $\mathbf{y}_1 = (0, 1, 0, 0, 1, 1, 0, \ldots)'$ for example. Note that all vectors are assumed to be *column vectors* and not *row vectors*, unless otherwise stated. The data matrix can be written in terms of the individual data vectors, or in terms of the data vectors for the variables, i.e.

$$\mathbf{X} = \begin{bmatrix} \mathbf{x}'_1 \\ \mathbf{x}'_2 \\ \vdots \\ \mathbf{x}'_r \\ \vdots \\ \mathbf{x}'_n \end{bmatrix} \quad \text{or} \quad \mathbf{X} = \begin{bmatrix} \mathbf{y}_1 & \mathbf{y}_2 & \cdots & \mathbf{y}_i & \cdots & \mathbf{y}_p \end{bmatrix}.$$

Multivariate data analysis attempts to make sense of these types of variables, together with the interactions between them. The various aims of multivariate data analysis can be summarized as follows:

Exploratory.   Suppose in our survey on holidays taken by young people, we had five thousand respondents and fifty variables covering demographics, education, previous holidays, habits and attitudes towards holidays. That is two hundred and fifty thousand pieces of information. Multivariate data analysis allows us to explore these data in order to extract information from them. For example, one of the simple pieces of information we would like to know is which are the most popular holiday destinations, and are these changing over time?

Simplification.   With such a lot of data, we need to simplify matters. This may involve choosing key variables and leaving out the rest. It may mean segmenting the respondents into meaningful groups, perhaps by age or destination. It may mean amalgamating variables or categories for a variable, for instance destinations might be grouped into continents, rather than individual countries.

Description and visualization.   Simple descriptions of the data are required, either pictorially or numerically. Histograms, pie charts, means, standard deviations, etc. give simple summaries of the data, but others are required to capture the multivariate nature of the data.

Interdependence.   With multivariate data, there are interdependencies among the variables. These need to be explored. Pairwise interdependence can be measured using correlation coefficients and other means, depending on the types of variable. However, we also need to measure and model the interdependence between many variables simultaneously.

Estimation and testing.   As with univariate data, population parameters can be estimated and hypotheses about them tested. The methods for the univariate case cannot be routinely used since they do not allow for the multivariate nature of the data.

Dimension reduction.   Our fifty variables give us fifty-dimensional data. We live in a three-dimensional world and it is easier for us to make sense of things in a space of a small number of dimensions. Some multivariate data analysis techniques reduce the dimensions of the data down to something more manageable. For instance, you cannot draw a fifty-dimensional scatterplot, but you can a two- or three-dimensional one.

Classification.   The human brain is continuously classifying, or otherwise we could not cope with the barrage of information that hits us through our eyes, ears, mouth, nose and skin. With so many variables and observations in our dataset, we might need to classify (group, cluster) the variables and/or observations in order to help assimilate the information extracted from the data.

Discrimination.   Suppose we look at two particular groups of the young holiday makers in our survey. The first is the group that are still in full-time education, and the second, the group that are in full-time employment. Which variables distinguish the two groups? For instance, do the destinations tend to differ between the two groups? Does the average amount spent per day differ? Discriminant analysis is a technique that highlights the difference between two or more groups. It can also be used to help place observations of unknown group into one of several possible groups. This is especially useful in the medical context where, based on a set of symptoms, a doctor will want to place a patient into one of two groups – the group of patients that have a particular disease, or the group of patients that exhibit similar symptoms, but do not actually have the disease.

Prediction.   In many instances we wish to predict future events. For a company that sells holidays to young people, they will want to know the future demand for holidays. Not only will the company wish to know the total number of holidays that they should provide in the next five years for young people, but in which countries, in which resorts, and the type of holidays required. Trends among young people might be changing very fast and the company needs to keep abreast of these and predict them into the future. Mathematical/statistical models of future holiday demand, fitted using the survey data, can help predict the future patterns.

Data mining.   With the increase in computer power and data storage capability, very large datasets now abound. Such datasets or databases are analysed using multivariate analysis techniques as with any smaller multivariate dataset. However, the handling of large datasets and the application of statistical techniques to them, can produce particular challenges.

1.2 Proximity data

There is one other type of data that we need to consider before proceeding. Some multivariate techniques use *proximity* data, for instance many multidimensional scaling techniques and clustering techniques. Proximity means 'nearness', and proximity measures the attempt to measure the nearness of one object to another. There are two types of proximity measure: *similarity* and *dissimilarity*. The similarity between two objects is a measure of 'how similar' they are, while dissimilarity is a measure of 'how dissimilar' they are. Consider two objects A and B. Then a dissimilarity measure, $\delta(A, B)$, between the two objects should satisfy the following conditions

$$\delta(A, B) \geq 0$$
$$\delta(A, A) = 0.$$

The first condition says that dissimilarities cannot be negative, and the second that an object cannot be dissimilar from itself. Also, we require that if two objects are very different, then their dissimilarity should be large, whilst if two objects are not very dissimilar, then the dissimilarity should be small.

Another condition we sometimes impose between any three objects, A, B and C, is that

$$\delta(A,B) \leq \delta(A,C) + \delta(C,B),$$

i.e. the dissimilarity measure satisfies the triangle inequality, and is thus a metric.

If we are measuring the similarity, $s(A, B)$, between two objects, then $s(A, B)$ should satisfy the conditions

$$s(A, B) \geq 0$$
$$s(A, A) = \text{max},$$

where max is the maximum similarity possible on the scale of measurement being used, and often this will be unity. An object cannot be more similar to another object than it is to itself.

In general, we will define the dissimilarity between the rth and sth objects as δ_{rs}, and the similarity between the rth and sth objects as s_{rs}. Transformations can be made from similarities to dissimilarities, and vice versa. Possible transformations are

$$\delta_{rs} = 1 - s_{rs}$$
$$\delta_{rs} = c - s_{rs}, \quad \text{for some constant } c$$
$$\delta_{rs} = \{2(1 - s_{rs})\}^{\frac{1}{2}}.$$

One advantage of dissimilarities over similarities is that dissimilarities are more 'distance-like', which means they lend themselves to the plotting of points in a space, where each point represents an object, and distances between points represent the dissimilarities (see Chapter 9 on multidimensional scaling).

1.2.1 Dissimilarities for quantitative data

Table 1.2 shows some dissimilarity measures for continuous or discrete quantitative data.

Table 1.2 Dissimilarity measures for quantitative data

Euclidean distance	$\delta_{rs} = \left\{ \sum_i (x_{ri} - x_{si})^2 \right\}^{\frac{1}{2}}$
Weighted Euclidean	$\delta_{rs} = \left\{ \sum_i w_i (x_{ri} - x_{si})^2 \right\}^{\frac{1}{2}}$
Mahalanobis distance	$\delta_{rs} = \{ (\mathbf{x}_r - \mathbf{x}_s)' \Sigma^{-1} (\mathbf{x}_r - \mathbf{x}_s) \}^{\frac{1}{2}}$
City block metric	$\delta_{rs} = \sum_i |x_{ri} - x_{si}|$
Minkowski metric	$\delta_{rs} = \left\{ \sum_i |x_{ri} - x_{si}|^\lambda \right\}^{\frac{1}{\lambda}} \quad \lambda \geq 1$
Angular separation	$\delta_{rs} = 1 - \dfrac{\sum_i x_{ri} x_{si}}{[\sum_i x_{ri}^2 \sum_i x_{si}^2]^{\frac{1}{2}}}$
Correlation	$\delta_{rs} = 1 - \dfrac{\sum_i (x_{ri} - \bar{x}_r)(x_{si} - \bar{x}_s)}{\left[\sum_i (x_{ri} - \bar{x}_r)^2 \sum_i (x_{si} - \bar{x}_s)^2 \right]^{\frac{1}{2}}}$

The Mahalanobis distance requires the covariance matrix for the variables (see Chapter 3). The dissimilarity measure based on the correlation coefficient is useful as a measure of dissimilarity between variables as opposed to dissimilarity between objects. Indeed, any dissimilarity measure for objects can be used to measure dissimilarity between variables, as long as it makes sense to do so and the measure also makes sense. The roles of objects and variables are reversed.

1.2.2 Binary data

When all the variables are binary, it is usual to construct a similarity coefficient, and then this can be transformed into a dissimilarity coefficient. For objects r and s a table is constructed as follows

		Object s		
		1	0	
Object r	1	a	b	$a + b$
	0	c	d	$c + d$
		$a + c$	$b + d$	$p = a + b$ $+ c + d$

The table shows the number of variables, a, for which both objects r and s scored '1', the number of variables, b, where object r scored '1' and object s scored '0', the number of variables, c, for which object r scored '0' and object s scored '1', and finally, the number, d, of variables where both objects scored '0'. The values of a, b, c and d are then used to calculate the similarity between objects r and s using a suitable similarity coefficient. Table 1.3 shows several of these.

A coefficient has to be chosen to suit the data and the problem, although it is often desirable to try more than one, hoping for robustness against choice. The similarities can be transformed into dissimilarities using one of the transformations mentioned earlier.

As an example, Table 1.4 shows some binary data for the author's pets, where a pet scores '1' if it possesses an attribute, and '0' otherwise. The similarity between the dog and the cat,

Table 1.3 Similarity coefficients for binary data

Simple matching coefficient	$s_{rs} = \dfrac{a+d}{a+b+c+d}$
Jaccard coefficient	$s_{rs} = \dfrac{a}{a+b+c}$
Russell, Rao	$s_{rs} = \dfrac{a}{a+b+c+d}$
Hamman	$s_{rs} = \dfrac{a-(b+c)+d}{a+b+c+d}$
Kulczynski	$s_{rs} = \dfrac{a}{b+c}$
Czekanowski, Sørensen, Dice	$s_{rs} = \dfrac{2a}{2a+b+c}$
Sokal, Sneath, Anderberg	$s_{rs} = \dfrac{a}{a+2(b+c)}$
Yule	$s_{rs} = \dfrac{ad-bc}{ad+bc}$

Table 1.4 Binary data for pets

Pet	Has teeth	Has fur/hair	Has feathers	Has two legs	Has four legs	Swims	Eats meat
Dog	1	1	0	0	1	1	1
Cat	1	1	0	0	1	0	1
Parrot	0	0	1	1	0	0	0
Ferret	1	1	0	0	1	1	1
Vulture	0	0	1	1	0	0	1
Frog	0	0	0	0	1	1	0

based on the simple matching coefficient, is $(4+2)/(4+1+0+2)=0.857$. Transforming this to a dissimilarity using $1-s_{rs}$, gives the dissimilarity between the dog and the cat as 0.143. The dissimilarities between all pairs of animals are placed in a matrix, $\mathbf{D}$, where the rsth element of $\mathbf{D}$ is equal to the dissimilarity between the rth and sth animal. Thus

$$\mathbf{D} = \begin{bmatrix} 0.0 & 0.143 & 1.000 & 0.000 & 0.857 & 0.429 \\ 0.143 & 0.0 & 0.857 & 0.143 & 0.714 & 0.571 \\ 1.000 & 0.857 & 0.0 & 1.000 & 0.143 & 0.571 \\ 0.000 & 0.143 & 1.000 & 0.0 & 0.857 & 0.429 \\ 0.857 & 0.714 & 0.143 & 0.857 & 0.0 & 0.714 \\ 0.429 & 0.571 & 0.571 & 0.429 & 0.714 & 0.0 \end{bmatrix}.$$

The set of dissimilarities, $\{\delta_{rs}\}$, can then be analysed using cluster analysis, multidimensional scaling, or other multivariate technique.

1.2.3 Nominal, ordinal data and mixed data

For nominal data, if objects r and s share the same category for variable i, then let $s_{rsi}=1$, and zero otherwise. Then a possible measure of similarity is $p^{-1}\sum_i s_{rsi}$. One possibility for ordinal data is to treat it as quantitative. However, other similarity measures can be used for nominal and ordinal data, see Cox and Cox (2000), or Gordon (1999).

For mixed data, where some of the variables are quantitative, some binary, and some categorical, Gower (1971) suggested a general similarity coefficient, s_{rs}, where

$$s_{rs} = \frac{\sum_{i=1}^{p} w_{rsi} s_{rsi}}{\sum_{i=1}^{p} w_{rsi}},$$

where s_{rsi} is the similarity between the rth and sth objects based on the ith variable alone, and w_{rsi} is unity if the rth and sth objects can be compared on the ith variable, and zero if they cannot. For instance, if object r has a missing value for variable i, then $w_{rsi} = 0$. Thus s_{rs} is an average over s_{rsi} for which there is a valued measured. Various measures for s_{rsi} are possible. Gower suggests the following for presence/absence binary data.

Object r	Object s	s_{rsi}	w_{rsi}
+	+	1	1
+	−	0	1
−	+	0	1
−	−	0	0

For nominal variables, Gower suggests $s_{rsi} = 1$ if objects r and s share the same categorization for variable i, and zero otherwise. For quantitative variables,

$$s_{rsi} = 1 - |x_{ri} - x_{si}|/R_i,$$

where R_i is the range of the observations for variable i.

1.3 Statistical software packages

Multivariate data analysis can only sensibly be carried out using a computer, and usually a statistical software package. The following are some of the packages that will carry out multivariate data analysis, some of which have a comprehensive repertoire, and others only cover a few techniques. The reader is invited to look at the following websites, which are in no particular order, for further details.

SAS	(http://www.sas.com)
SPSS	(http://www.spss.com)
SYSTAT	(http://www.systat.com)
STATGRAPHICS	(http://www.statgraphics.com)
MINITAB	(http://www.minitab.com)
JMP	(http://www.jmp.com)
S-Plus	(http://www.insightful.com)
R	(http://cran.r-project.org)

1.4 The rest of the book

This book introduces the reader to a wide range of multivariate data analysis techniques. It cannot cover everything of course, but references to books for more detailed coverage are given in the text. Chapter 2 reinforces the reader's knowledge of matrix algebra. Chapter 3 covers basic multivariate statistics, and proceeds as would an introductory course in univariate statistics, but in the multivariate setting. The following chapters then cover the various multivariate techniques, with Chapters 4–9 dealing with techniques that are generally descriptive in nature, while Chapters 10–17 are model based. Finally, Chapter 18 gives a brief introduction to data mining.

1.5 Exercises

1. Look at the following websites and download data that you find interesting. These datasets can be used for practising multivariate data analysis techniques later.

   http://lib.stat.cmu.edu/DASL/ (The Data and Story Library)
   http://www.stat.ucla.edu/data/
   http://biostat.mc.vanderbilt.edu/twiki/bin/view/Main/DataSets
   http://www.ics.uci.edu/~mlearn/MLSummary.html
   http://www.statserv.com/datasets.html
   http://www.statsci.org/data/ (OzDASL)

   Datasets can be found in the books

   Data, D.F. Andrews and A.M. Herzberg, Springer (1985)
   A Handbook of Small Data Sets, D.J. Hand, F. Daly, K. McConway,
       D. Lunn, and E. Ostrowski, Chapman & Hall/CRC (1994)

   The data from these and other books can also be found at various websites.

   Various statistical resources can be found at the sites

   http://www.stat.ufl.edu/vlib/statistics.html
   http://www.amstat.org/
   http://www.rss.org.uk/

   and there are many others that can be found using a web search engine.

2. For the binary data on pets in Table 1.4, calculate a matrix of dissimilarities using the Jaccard coefficient, together with the transformation, $\delta_{rs} = 1 - s_{rs}$.

3. Calculate the similarity between respondent number 1 and respondent number 2 for the holiday destination data of Table 1.1, using Gower's general similarity coefficient. (Take the range of x_2 to be 16, and that for x_3 as 200.) Do the same for respondents 1 and 6.

2

Matrix algebra

Some knowledge of matrix algebra is necessary for a mathematical study of multivariate data analysis, and it is assumed that the reader has some familiarity with such. This chapter will cover that needed for the rest of the book. Nothing is proved, only various results stated, and sometimes illustrated. The reader may also find it instructive to practice matrix manipulations, either by hand (usually restricted to 2×2 matrices), or by using a mathematical or statistical package that will carry out interactive matrix algebra, for instance, MATLAB, MAPLE, MATHEMATICA, MINITAB, R, or S-Plus.

2.1 Vectors and matrices

A vector is a column of p elements, and will be written in lower case bold font, such as $\mathbf{x}$, $\mathbf{y}$, $\mathbf{a}$ or $\mathbf{b}$. The elements are usually numbers. Each element represents a 'dimension', and we say that the vector is a p-dimensional vector. The following are two vectors, $\mathbf{x}$ and $\boldsymbol{\alpha}$:

$$\mathbf{x} = \begin{pmatrix} 3.7 \\ 0.6 \\ 9.7 \end{pmatrix} \quad \boldsymbol{\alpha} = \begin{pmatrix} \alpha_1 \\ \alpha_2 \\ \vdots \\ \alpha_p \end{pmatrix}.$$

The element in the ith column of $\mathbf{x}$ is denoted by x_i, or $[\mathbf{x}]_i$. These vectors, so defined, are *column vectors*. In a similar manner, *row vectors* are defined as rows of elements. Usually, when we refer to a vector, we will mean a column vector, unless otherwise stated, or it is clear from the context.

Matrices are two-dimensional arrays of elements, and are usually referred to by upper case, bold font letters. We say a matrix is a $p \times q$ matrix, if it has p rows and q columns. Alternatively, we say the matrix is of order $p \times q$. The following are two matrices, $\mathbf{X}$ and $\boldsymbol{\Sigma}$:

$$\mathbf{X} = \begin{pmatrix} 3.7 & 4.2 & 5.0 \\ 0.6 & 3.9 & 2.1 \\ 9.7 & 0.0 & 6.6 \end{pmatrix} \quad \boldsymbol{\Sigma} = \begin{pmatrix} \sigma_{11} & \sigma_{12} & \cdots & \sigma_{1p} \\ \sigma_{21} & \sigma_{22} & \cdots & \sigma_{2p} \\ \vdots & \vdots & \ddots & \vdots \\ \sigma_{p1} & \sigma_{p2} & \cdots & \sigma_{pp} \end{pmatrix}.$$

The (i,j)th element of a matrix is the element placed in the ith row, and jth column. For example, the $(2,1)$th element of **X** is 0.6, and the $(3,4)$th element of **Σ** above is σ_{34}. We sometimes write a matrix as $[\sigma_{ij}]$, and the (i,j)th element of the matrix as $[\mathbf{\Sigma}]_{ij}$.

Matrices are sometimes viewed as a matrix of column vectors, or a matrix of row vectors. For instance, if

$$\mathbf{X} = \begin{pmatrix} 1 & 3 & 8 \\ 4 & 4 & 2 \\ 0 & 1 & 9 \\ 3 & 4 & 5 \end{pmatrix},$$

then **X** can be viewed as a matrix consisting of the column vectors

$$\begin{pmatrix} 1 \\ 4 \\ 0 \\ 3 \end{pmatrix}, \begin{pmatrix} 3 \\ 4 \\ 1 \\ 4 \end{pmatrix}, \begin{pmatrix} 8 \\ 2 \\ 9 \\ 5 \end{pmatrix},$$

or a matrix of row vectors

$$\begin{pmatrix} 1 & 3 & 8 \end{pmatrix}, \begin{pmatrix} 4 & 4 & 2 \end{pmatrix}, \begin{pmatrix} 0 & 1 & 9 \end{pmatrix}, \begin{pmatrix} 3 & 4 & 5 \end{pmatrix}.$$

2.2 Operations on vectors and matrices

In this section, we review some of the basic operations on matrices and vectors. We will use the following matrices and vectors, **A**, **B**, **C**, **d**, **e**, for illustration:

$$\mathbf{A} = \begin{pmatrix} 4 & 2 & 1 \\ 2 & 6 & 3 \\ 1 & 3 & 2 \end{pmatrix} \quad \mathbf{B} = \begin{pmatrix} -1 & 0 & 7 \\ 5 & 2 & -3 \\ -2 & 1 & 1 \end{pmatrix} \quad \mathbf{C} = \begin{pmatrix} 2 & 1 \\ 5 & 1 \\ 1 & -2 \end{pmatrix}$$

$$\mathbf{d} = \begin{pmatrix} 1 \\ 1 \\ 7 \end{pmatrix} \quad \mathbf{e} = \begin{pmatrix} -3 \\ 3 \\ -3 \end{pmatrix}.$$

Square matrix. A square matrix is one which has the same number of rows as columns. Matrices **A** and **B** are square.

Transpose of a matrix. The transpose of a matrix, **A**, is a new matrix, **A′**, where $[\mathbf{A'}]_{ij} = [\mathbf{A}]_{ji}$ (i.e. to form **A′**, write the rows of **A** as columns for **A′**, or vice versa). Thus

$$\mathbf{C'} = \begin{pmatrix} 2 & 5 & 1 \\ 1 & 1 & -2 \end{pmatrix}.$$

Transpose also applies to vectors, so that the transpose of a column vector is a row vector, and vice versa.

Matrix addition.   Two matrices of the same order can be added together, by adding corresponding elements. For example

$$\mathbf{A} + \mathbf{B} = \begin{pmatrix} 3 & 2 & 8 \\ 7 & 8 & 0 \\ -1 & 4 & 3 \end{pmatrix}.$$

Matrix subtraction.   A matrix can be subtracted from another matrix of the same order. For example

$$\mathbf{A} - \mathbf{B} = \begin{pmatrix} 5 & 2 & -6 \\ -3 & 4 & 6 \\ 3 & 2 & 1 \end{pmatrix}.$$

Scalar multiplication.   Multiplication of a matrix by a scalar occurs when each element of the matrix is multiplied by the scalar. Thus

$$6\mathbf{A} = \begin{pmatrix} 24 & 12 & 6 \\ 12 & 36 & 18 \\ 6 & 18 & 12 \end{pmatrix}.$$

Inner product of two vectors.   The inner product, $\mathbf{x}'\mathbf{y}$, or $\langle \mathbf{x}, \mathbf{y} \rangle$, of two vectors, is the scalar quantity, $\sum_i x_i y_i$. Thus

$$\mathbf{d}'\mathbf{e} = \begin{pmatrix} 1 & 1 & 7 \end{pmatrix} \begin{pmatrix} -3 \\ 3 \\ -3 \end{pmatrix} = -21.$$

Norm of a vector.   The norm, $||\mathbf{x}||$, of a vector $\mathbf{x}$ is the quantity $\sqrt{\mathbf{x}'\mathbf{x}}$ which measures its length. Vector $\mathbf{x}$ can be standardized to have unit length, i.e. $\mathbf{x}/\sqrt{\mathbf{x}'\mathbf{x}}$. For example, the length of $\mathbf{d}$ is $\sqrt{51}$, and the vector

$$\begin{pmatrix} 1/\sqrt{51} \\ 1/\sqrt{51} \\ 7/\sqrt{51} \end{pmatrix}$$

is vector $\mathbf{d}$ standardized.

Matrix multiplication.   A $p \times q$ matrix, $\mathbf{A}$, can be multiplied by a $q \times r$ matrix, $\mathbf{B}$ (note that the number of columns of $\mathbf{A}$ must be equal to the number of rows of $\mathbf{B}$). Let $\mathbf{A}$ be written as a matrix of column vectors, so that $\mathbf{A} = [\mathbf{a}_1, \mathbf{a}_2, \ldots, \mathbf{a}_q]$, and $\mathbf{B}$ also as a matrix of column vectors, $\mathbf{B} = [\mathbf{b}_1, \mathbf{b}_2, \ldots, \mathbf{b}_r]$. Then the ijth element of the product $\mathbf{AB}$ is $\mathbf{a}_i' \mathbf{b}_j$. Thus

$$\mathbf{AB} = \begin{pmatrix} 4 & 5 & 23 \\ 22 & 15 & -1 \\ 10 & 8 & 0 \end{pmatrix}.$$

Trace of a matrix.   The trace, tr $\mathbf{X}$, of a square matrix $\mathbf{X}$, is the sum of its diagonal elements, $\sum_i x_{ii}$. Thus tr $\mathbf{A} = 12$. Two properties of tr are: $\text{tr}(\mathbf{AB}) = \text{tr}(\mathbf{BA})$, and $\text{tr}(\mathbf{A} + \mathbf{B}) = \text{tr }\mathbf{A} + \text{tr }\mathbf{B}$.

Determinant of a matrix.   The determinant, $|\mathbf{A}|$, of a 2 $\times$ 2 matrix, $\mathbf{A}$, is the scalar quantity,

$$|\mathbf{A}| = \begin{vmatrix} a_{11} & a_{12} \\ a_{21} & a_{22} \end{vmatrix} = a_{11}a_{22} - a_{12}a_{21}.$$

The determinant of a 3×3 matrix is

$$\begin{vmatrix} a_{11} & a_{12} & a_{13} \\ a_{21} & a_{22} & a_{23} \\ a_{31} & a_{32} & a_{33} \end{vmatrix} = a_{11}a_{22}a_{33} - a_{11}a_{23}a_{32} + a_{12}a_{23}a_{31}$$

$$- a_{12}a_{21}a_{33} + a_{13}a_{21}a_{32} - a_{13}a_{22}a_{31}.$$

For matrices of higher order, rather than writing the determinant as a sum of terms with a particular pattern of suffices for a_{ij}, it is easier to define it in terms of cofactors. However, before defining cofactor, we define the minors for matrix $\mathbf{A}$. The minor M_{ij} of a_{ij} is the determinant of the submatrix obtained from $\mathbf{A}$ by deleting the ith row and jth column. The cofactor, A_{ij}, of a_{ij} is then defined as $(-1)^{(i+j)}M_{ij}$. The determinant of $\mathbf{A}$ is then

$$|\mathbf{A}| = a_{i1}A_{i1} + a_{i2}A_{i2} + \cdots + a_{in}A_{in},$$

where i can be chosen as any one of the n rows. Alternatively, expanding by a column,

$$|\mathbf{A}| = a_{1j}A_{1j} + a_{2j}A_{2j} + \cdots + a_{nj}A_{nj},$$

where j can be chosen as any one of the n columns. Thus for example, the cofactors of a_{11}, a_{12}, and a_{13} of $\mathbf{A}$ are

$$A_{11} = (-1)^{(1+1)} \begin{vmatrix} 6 & 3 \\ 3 & 2 \end{vmatrix} = (1)(6 \times 2 - 3 \times 3) = 3$$

$$A_{12} = (-1)^{(1+2)} \begin{vmatrix} 2 & 3 \\ 1 & 2 \end{vmatrix} = (-1)(2 \times 2 - 1 \times 3) = -1$$

$$A_{13} = (-1)^{(1+3)} \begin{vmatrix} 2 & 6 \\ 1 & 3 \end{vmatrix} = (1)(2 \times 3 - 1 \times 6) = 0.$$

Thus $|\mathbf{A}| = 4 \times 3 + 2 \times (-1) + 1 \times 0 = 10$. Note that determinants are only defined for square matrices.

Singular matrices.   The matrix $\mathbf{A}$ is singular if $|\mathbf{A}| = 0$. The matrix $\mathbf{A}$ is non-singular if $|\mathbf{A}| \neq 0$.

Inverse of a matrix.   The inverse of a square matrix, $\mathbf{A}$, is that matrix, $\mathbf{A}^{-1}$, which satisfies

$$\mathbf{A}\mathbf{A}^{-1} = \mathbf{A}^{-1}\mathbf{A} = \mathbf{I}.$$

The inverse, $\mathbf{A}^{-1}$, only exists if $\mathbf{A}$ is non-singular.

Partitioned matrix.   A partitioned matrix is one that is written in terms of *submatrices*. For example

$$\mathbf{A} = \begin{pmatrix} \mathbf{A}_{11} & \mathbf{A}_{12} \\ \mathbf{A}_{21} & \mathbf{A}_{22} \end{pmatrix} = \left(\begin{array}{cc|c} 4 & 2 & 1 \\ 2 & 6 & 3 \\ \hline 1 & 3 & 2 \end{array} \right),$$

where

$$\mathbf{A}_{11} = \begin{pmatrix} 4 & 2 \\ 2 & 6 \end{pmatrix} \quad \mathbf{A}_{12} = \begin{pmatrix} 1 \\ 3 \end{pmatrix} \quad \mathbf{A}_{21} = (1 \quad 3) \quad \mathbf{A}_{22} = (2).$$

The following are some properties of $n \times n$ determinants.

$$|\mathbf{A}'| = |\mathbf{A}|$$

$$k|\mathbf{A}| = k^n|\mathbf{A}| \quad \text{where } k \text{ is a constant}$$

$$|\mathbf{A}^{-1}| = \frac{1}{|\mathbf{A}|}$$

$$|\mathbf{AB}| = |\mathbf{BA}|$$

$$|\mathbf{AB}| = |\mathbf{B}|\,|\mathbf{A}|$$

$$\begin{vmatrix} \mathbf{A}_{11} & \mathbf{A}_{12} \\ \mathbf{A}_{21} & \mathbf{A}_{22} \end{vmatrix} = |\mathbf{A}_{11}||\mathbf{A}_{22} - \mathbf{A}_{21}\mathbf{A}_{11}^{-1}\mathbf{A}_{12}| = |\mathbf{A}_{22}||\mathbf{A}_{11} - \mathbf{A}_{12}\mathbf{A}_{22}^{-1}\mathbf{A}_{21}|.$$

The following are some properties of inverses.

$$(k\mathbf{A})^{-1} = k^{-1}\mathbf{A}^{-1}$$

$$(\mathbf{AB})^{-1} = \mathbf{B}^{-1}\mathbf{A}^{-1}.$$

The transposed matrix of all the cofactors of $\mathbf{A}$ is the *adjoint*, adj $\mathbf{A}$, of $\mathbf{A}$. For example

$$\text{adj } \mathbf{A} = \begin{pmatrix} A_{11} & A_{21} & A_{31} \\ A_{12} & A_{22} & A_{32} \\ A_{13} & A_{23} & A_{33} \end{pmatrix} = \begin{pmatrix} 3 & -1 & 0 \\ -1 & 7 & -10 \\ 0 & -10 & 20 \end{pmatrix}.$$

Then the inverse, $\mathbf{A}^{-1}$, can be calculated as

$$\mathbf{A}^{-1} = \frac{1}{|\mathbf{A}|}\text{adj } \mathbf{A}.$$

Thus

$$\mathbf{A}^{-1} = \frac{1}{10}\begin{pmatrix} 3 & -1 & 0 \\ -1 & 7 & -10 \\ 0 & -10 & 20 \end{pmatrix} = \begin{pmatrix} 0.3 & -0.1 & 0 \\ -0.1 & 0.7 & -1 \\ 0 & -1 & 2 \end{pmatrix}.$$

Rank of a matrix. The rank, $R(\mathbf{A})$, of the $p \times q$ matrix, $\mathbf{A}$, is the maximum number of linearly independent rows or columns of the matrix. A set of row (column) vectors is linearly independent if it is impossible to write any one of them as a linear combination of the rest. It can be shown that $R(\mathbf{A}) \leq \min(p, q)$. As an example, the rank of $\mathbf{A}$ is three, whilst the rank of the following matrix is two,

$$\begin{pmatrix} 2 & 1 & 0 \\ 1 & 1 & 4 \\ 4 & 3 & 8 \end{pmatrix},$$

because there are only two linearly independent rows in the matrix, since the third row can be obtained as a linear sum of the other two rows: $(4, 3, 8) = (2, 1, 0) + 2(1, 1, 4)$.

2.3 Types of matrices

There are some special types of matrices that we will need.

Diagonal matrix. The 'diagonal' of a $p \times p$ square matrix, $\mathbf{A}$, is the line of elements, starting from element a_{11} and ending at a_{pp}. If all other elements of the matrix are zero, then it is a diagonal matrix. For example

$$\begin{pmatrix} 6 & 0 & 0 & \cdots & 0 \\ 0 & 3 & 0 & \cdots & 0 \\ \vdots & \vdots & \vdots & \ddots & \vdots \\ 0 & 0 & 0 & \cdots & 4 \end{pmatrix},$$

and is written as diag$(6, 3, \ldots, 4)$.

Symmetric matrix. A symmetric matrix is a square matrix such that its ijth element is equal to its jith element. If $\sigma_{ij} = \sigma_{ji}$ in the matrix $\mathbf{\Sigma}$ above, then $\mathbf{\Sigma}$ is a symmetric matrix. On the other hand, if $\sigma_{ij} \neq \sigma_{ji}$, then $\mathbf{\Sigma}$ would be an *asymmetric* matrix.

Identity matrix. A diagonal matrix, $\mathbf{I}$ or $\mathbf{I}_p$, that has ones on its diagonal, is an identity matrix,

$$\mathbf{I} = \begin{pmatrix} 1 & 0 & 0 & \cdots & 0 \\ 0 & 1 & 0 & \cdots & 0 \\ \vdots & \vdots & \vdots & \ddots & \vdots \\ 0 & 0 & 0 & \cdots & 1 \end{pmatrix}.$$

The vector 1. The vector $\mathbf{1}$ or $\mathbf{1}_p$ has each of its elements equal to unity. For example,

$$\mathbf{1}_3 = \begin{pmatrix} 1 \\ 1 \\ 1 \end{pmatrix}.$$

The matrix of ones. The matrix of ones has all its elements equal to unity. For example

$$\mathbf{E} = \begin{pmatrix} 1 & 1 & 1 \\ 1 & 1 & 1 \\ 1 & 1 & 1 \end{pmatrix}.$$

Note that $\mathbf{E} = \mathbf{1}_3 \mathbf{1}_3'$, where $\mathbf{1}_3'$ is a row vector of ones.

Zero vector. The vector $\mathbf{0}$ has each of its elements equal to zero.

The centring matrix. The centring matrix, $\mathbf{H}$, is defined as

$$\mathbf{H} = \begin{pmatrix} 1 - \frac{1}{p} & -\frac{1}{p} & -\frac{1}{p} & \cdots & -\frac{1}{p} \\ -\frac{1}{p} & 1 - \frac{1}{p} & -\frac{1}{p} & \cdots & -\frac{1}{p} \\ -\frac{1}{p} & -\frac{1}{p} & 1 - \frac{1}{p} & \cdots & -\frac{1}{p} \\ \vdots & \ddots & \ddots & \ddots & \vdots \\ -\frac{1}{p} & -\frac{1}{p} & -\frac{1}{p} & \cdots & 1 - \frac{1}{p} \end{pmatrix}.$$

The centring matrix is used to mean correct rows and columns of matrices. It can be written as

$$\mathbf{H} = \mathbf{I} - \frac{1}{p}\mathbf{11'}.$$

Orthogonal vectors.   Two vectors, $\mathbf{x}$ and $\mathbf{y}$, are orthogonal if $\mathbf{x'y} = 0$. This implies that the angle between the vectors is 90°.

Orthogonal matrix.   A square matrix, $\mathbf{A}$, is orthogonal if $\mathbf{A'A} = \mathbf{AA'} = \mathbf{I}$. Thus all the pairs of column (row) vectors comprising the matrix $\mathbf{A}$ are orthogonal to each other. It can be shown that for an orthogonal matrix, $\mathbf{A}^{-1} = \mathbf{A'}$ and $|\mathbf{A}| = \pm 1$.

2.4 Eigenvalues and eigenvectors

Eigenvalues and eigenvectors are defined for square matrices, and play a key role in multivariate data analysis. We introduce them by considering the following equation, where $\mathbf{A}$ is a known matrix,

$$\mathbf{Ax} = \lambda \mathbf{x}. \tag{2.1}$$

The scalar quantities, λ_i, that solve this equation are called eigenvalues, and the corresponding vectors, $\mathbf{x}_i$, are the eigenvectors. Rewrite equation (2.1) as

$$(\mathbf{A} - \lambda \mathbf{I})\mathbf{x} = \mathbf{0}. \tag{2.2}$$

For equation (2.2) to have a non-trivial solution (i.e. $\mathbf{x} \neq \mathbf{0}$),

$$|\mathbf{A} - \lambda \mathbf{I}| = 0,$$

and the solution of this equation gives us the eigenvalues of $\mathbf{A}$. For our matrix $\mathbf{A}$,

$$|\mathbf{A} - \lambda \mathbf{I}| = \begin{vmatrix} 4 - \lambda & 2 & 1 \\ 2 & 6 - \lambda & 3 \\ 1 & 3 & 2 - \lambda \end{vmatrix} = 0.$$

Expanding the determinant gives the cubic equation

$$\lambda^3 - 12\lambda^2 + 30\lambda - 10 = 0.$$

The roots of this are $\lambda_1 = 8.67$, $\lambda_2 = 2.03$ and $\lambda_3 = 0.39$ to 2 d.p. These are the three eigenvalues of $\mathbf{A}$. To find the eigenvector corresponding to λ_1, we now solve

$$\begin{pmatrix} 4 & 2 & 1 \\ 2 & 6 & 3 \\ 1 & 3 & 2 \end{pmatrix} \begin{pmatrix} x_1 \\ x_2 \\ x_3 \end{pmatrix} = 8.67 \begin{pmatrix} x_1 \\ x_2 \\ x_3 \end{pmatrix}.$$

Expanding this gives the simultaneous equations

$$-4.67x_1 + 2x_2 + x_3 = 0$$
$$2x_1 - 2.67x_2 + 3x_3 = 0$$
$$x_1 + 3x_2 - 6.67x_3 = 0.$$

There is not a unique solution to these equations, since they are not linearly independent. We can solve for x_1 and x_2 in terms of x_3 though. This gives

$$x_1 = 1.02x_3, \quad x_2 = 1.89x_3.$$

Thus the eigenvector, x_1, corresponding to λ_1, is

$$\begin{pmatrix} 1.02x_3 \\ 1.89x_3 \\ x_3 \end{pmatrix}.$$

This has length $2.37x_3$. Eigenvectors can be of arbitrary length, and so we usually standardize them to have unit length. Thus the standardized eigenvector, $\mathbf{x}_1$, is

$$\begin{pmatrix} 0.43 \\ 0.80 \\ 0.42 \end{pmatrix}.$$

The other two eigenvectors can be found in a similar manner, and are

$$\begin{pmatrix} 0.90 \\ -0.37 \\ -0.22 \end{pmatrix} \begin{pmatrix} -0.02 \\ 0.48 \\ -0.88 \end{pmatrix}.$$

The eigenvectors defined so far are *right* eigenvectors, and are usually just called eigenvectors. Another set of eigenvectors, $\mathbf{y}$, called the *left* eigenvectors, are defined by

$$\mathbf{y}'\mathbf{A} = \lambda\mathbf{y}'.$$

Some properties of eigenvalues. It can be shown that

$$|\mathbf{A}| = \prod_i \lambda_i$$

$$\mathrm{tr}(\mathbf{A}) = \sum_i \lambda_i.$$

The eigenvalues of a matrix can be real or complex. However, all the eigenvalues of a symmetric matrix are real.

2.4.1 Spectral decomposition

A symmetric matrix, $\mathbf{A}$, can be written in the following form

$$\mathbf{A} = \mathbf{V}\mathbf{\Lambda}\mathbf{V}' = \sum_i \lambda_i \mathbf{v}_i'\mathbf{v}_i,$$

where $\mathbf{\Lambda}$ is the diagonal matrix, consisting of the eigenvalues of $\mathbf{A}$, and $\mathbf{V}$ is the matrix formed by the eigenvectors of $\mathbf{A}$. The matrix $\mathbf{V}$ is orthogonal, so $\mathbf{V}'\mathbf{V} = \mathbf{I}$. For example, using our derived eigenvalues and eigenvectors of $\mathbf{A}$,

$$\begin{pmatrix} 4 & 2 & 1 \\ 2 & 6 & 3 \\ 1 & 3 & 2 \end{pmatrix} = \begin{pmatrix} 0.43 & 0.90 & -0.02 \\ 0.80 & -0.37 & 0.48 \\ 0.42 & -0.22 & -0.88 \end{pmatrix} \begin{pmatrix} 8.67 & 0 & 0 \\ 0 & 2.03 & 0 \\ 0 & 0 & 0.39 \end{pmatrix}$$

$$\times \begin{pmatrix} 0.43 & 0.80 & 0.42 \\ 0.90 & -0.37 & -0.22 \\ -0.02 & 0.48 & -0.88 \end{pmatrix}.$$

Fractional and other powers of symmetric matrices are easily found using the spectral decomposition. For example, $\mathbf{A}^{1/2}$ is the matrix which multiplied by itself is equal to $\mathbf{A}$, and it is easily seen that

$$\mathbf{A}^{1/2} = \mathbf{V}\mathbf{\Lambda}^{1/2}\mathbf{V}',$$

where $\mathbf{\Lambda}^{1/2}$ is the diagonal matrix consisting of the square roots of the eigenvalues of $\mathbf{A}$. Similarly $\mathbf{A}^{-1} = \mathbf{V}\mathbf{\Lambda}^{-1}\mathbf{V}'$, where $\mathbf{\Lambda}^{-1}$ is the diagonal matrix with the elements λ_i^{-1} down the diagonal. Thus for our matrix $\mathbf{A}$,

$$\mathbf{A}^{1/2} = \begin{pmatrix} 1.94 & 0.43 & 0.201 \\ 0.43 & 2.25 & 0.87 \\ 0.20 & 0.87 & 1.10 \end{pmatrix} \quad \mathbf{A}^{-1} = \begin{pmatrix} 0.3 & -0.1 & 0.0 \\ -0.1 & 0.7 & -1.0 \\ 0.0 & -1.0 & 2.0 \end{pmatrix}.$$

2.4.2 Singular valued decomposition

A similar decomposition to that of the spectral decomposition, but which applies to any matrix, is the singular value decomposition (SVD). A $p \times q$ matrix, $\mathbf{A}$, which is of rank r, can be written as

$$\mathbf{A} = \mathbf{U}\mathbf{\Gamma}\mathbf{V}',$$

where $\mathbf{U}$ is a $p \times r$ matrix, $\mathbf{\Gamma} = \text{diag}(\gamma_1, \ldots, \gamma_r)$, is an $r \times r$ diagonal matrix, whose diagonal elements are called the singular values of $\mathbf{A}$, and are such that, $\gamma_1 \geq \gamma_2 \geq \ldots \geq \gamma_r > 0$, and $\mathbf{V}$ is an $r \times q$ matrix. The matrices $\mathbf{U}$ and $\mathbf{V}$ are both orthogonal, and it can be shown that $\mathbf{U}$ consists of the eigenvectors of $\mathbf{A}\mathbf{A}'$, $\mathbf{V}$ consists of the eigenvectors of $\mathbf{A}'\mathbf{A}$, and $\mathbf{\Gamma} = \text{diag}(\sqrt{\lambda_1}, \sqrt{\lambda_2}, \ldots, \sqrt{\lambda_r})$, where $\{\mathbf{\Lambda}_i\}$ are the eigenvalues of $\mathbf{A}\mathbf{A}'$ (or $\mathbf{A}'\mathbf{A}$). For example

$$\begin{pmatrix} -1 & 0 & 7 \\ 5 & 2 & -3 \\ -2 & 1 & 1 \end{pmatrix} = \begin{pmatrix} -0.76 & 0.65 & -0.03 \\ 0.62 & 0.74 & 0.28 \\ -0.20 & -0.19 & 0.96 \end{pmatrix} \begin{pmatrix} 8.58 & 0 & 0 \\ 0 & 4.23 & 0 \\ 0 & 0 & 1.60 \end{pmatrix}$$

$$\times \begin{pmatrix} -0.50 & -0.81 & 0.32 \\ -0.12 & -0.30 & -0.95 \\ 0.86 & -0.51 & 0.05 \end{pmatrix}.$$

2.5 Quadratic forms

A quadratic form, $Q(\mathbf{x})$, is defined as

$$Q(\mathbf{x}) = \sum_{i=1}^{p} \sum_{j-1}^{p} a_{ij} x_i x_j,$$

where $a_{ij} = a_{ji}$, and $x_1, \ldots, x_p$ are p real variables. Placing the $\{a_{ij}\}$ in the symmetric matrix $\mathbf{A}$, $Q(\mathbf{x})$ can be written as

$$Q(\mathbf{X}) = \mathbf{x}' \mathbf{A} \mathbf{x}.$$

The matrix $\mathbf{A}$ is called positive definite if $Q(\mathbf{x}) > 0$ for all $\mathbf{x} \neq \mathbf{0}$, and positive semi-definite if $Q(\mathbf{x}) \geq 0$ for all $\mathbf{x} \neq \mathbf{0}$. Negative definite and negative semi-definite are defined in like manner. For example, the quadratic form using our matrix $\mathbf{A}$ is

$$Q(\mathbf{x}) = 4x_1^2 + 6x_2^2 + 2x_3^2 + 4x_1 x_2 + 2x_1 x_3 + 6x_2 x_3.$$

Now, $Q(\mathbf{x})$ can be written as

$$Q(\mathbf{x}) = 8.67(0.43x_1 + 0.80x_2 + 0.42x_3)^2 + 2.03(0.90x_1 - 0.37x_2 - 0.22x_3)^2$$
$$+ 0.39(-0.02x_1 + 0.48x_2 - 0.88x_3)^2.$$

Notice the use of the spectral decomposition of $\mathbf{A}$ here. Because all three eigenvalues are positive, and the squared terms have to be non-negative, $Q(\mathbf{x}) > 0$, for all $\mathbf{x} \neq \mathbf{0}$. Hence $\mathbf{A}$ is positive definite. Generally, $\mathbf{A}$ is positive (negative) definite if its eigenvalues, λ_i, are all positive (negative). If all the eigenvalues are non-negative (non-positive), i.e. some can be zero, then $\mathbf{A}$ is positive semi-definite (negative semi-definite).

2.6 Vector and matrix differentiation

Let $f(\mathbf{x})$ be a scalar function of vector $\mathbf{x}$. Then the derivative of $f(\mathbf{x})$ with respect to $\mathbf{x}$ is the vector

$$\frac{\partial f(\mathbf{x})}{\partial \mathbf{x}} = \left(\frac{\partial f(\mathbf{x})}{\partial x_i} \right).$$

For example, consider the derivative of $\mathbf{a}'\mathbf{x}$. Let $\mathbf{a}' = (a_1, a_2, \ldots, a_p)$, and $\mathbf{x}' = (x_1, x_2, \ldots, x_p)$. Then

$$\mathbf{a}'\mathbf{x} = a_1 x_1 + a_2 x_2 + \cdots + a_p x_p.$$

Differentiating with respect to x_i,

$$\frac{\partial \mathbf{a}'\mathbf{x}}{\partial x_i} = a_i.$$

Placing these partial derivatives into a vector gives

$$
\begin{pmatrix} \dfrac{\partial \mathbf{a}'\mathbf{x}}{\partial x_1} \\[2mm] \dfrac{\partial \mathbf{a}'\mathbf{x}}{\partial x_2} \\[2mm] \vdots \\[2mm] \dfrac{\partial \mathbf{a}'\mathbf{x}}{\partial x_p} \end{pmatrix} = \begin{pmatrix} a_1 \\ a_2 \\ \vdots \\ a_p \end{pmatrix} = \mathbf{a}.
$$

The following results will be useful in later chapters

$$
\frac{\partial \mathbf{x}'\mathbf{x}}{\partial \mathbf{x}} = 2\mathbf{x}
$$

$$
\frac{\partial \mathbf{x}'\mathbf{A}\mathbf{x}}{\partial \mathbf{x}} = 2\mathbf{A}\mathbf{x} \quad \text{(if } \mathbf{A} \text{ symmetric)}
$$

$$
\frac{\partial \mathbf{x}'\mathbf{A}\mathbf{x}}{\partial \mathbf{x}} = (\mathbf{A} + \mathbf{A}')\mathbf{x} \quad \text{(if } \mathbf{A} \text{ not symmetric)}.
$$

The differentiation with respect to a matrix is similarly defined:

$$
\frac{\partial f(\mathbf{X})}{\partial \mathbf{X}} = \left(\frac{\partial f(\mathbf{X})}{\partial X_{ij}} \right).
$$

2.7 Lagrange multipliers

Lagrange multipliers help in the minimization or maximization of a function of several variables, when there are equality constraints on the variables. We illustrate how they are used with the following simple example:

$$
\text{maximize } xy, \quad \text{subject to } y = 3 - 2x.
$$

First write the constraint as

$$
y + 2x - 3 = 0,
$$

and then we maximize the function, L, given by

$$
L = xy - \lambda(y + 2x - 3).
$$

The λ is a constant – the Lagrange multiplier. Differentiating L with respect to x, y and λ, and equating to zero, gives

$$
\frac{\partial L}{\partial x} = y - 2\lambda = 0
$$

$$
\frac{\partial L}{\partial y} = x - \lambda = 0
$$

$$
\frac{\partial L}{\partial \lambda} = y + 2x - 3 = 0.
$$

The first two equations give $y = 2\lambda$, and $x = \lambda$. The last equation simply gives the imposed constraint. Substituting for x and y in the constraint gives $2\lambda + 2\lambda - 3 = 0$, and hence $\lambda = 3/4$. Thus a stationary point exists at $x = 3/4$, $y = 3/2$, and we can check that this is a maximum, rather than a minimum. The maximum value of the function is $9/8$.

This is obviously a trivial example as we could have written the function in terms of x only, i.e. $x(3 - 2x)$, and maximized this. However, with complicated functions and constraints, it is more convenient to use Lagrange multipliers, using one for each constraint.

2.8 Exercises

1. Let vectors $\mathbf{x}$ and $\mathbf{y}$ be given by

$$\mathbf{x} = \begin{pmatrix} 1 \\ 8 \end{pmatrix} \quad \mathbf{y} = \begin{pmatrix} -3 \\ 2 \end{pmatrix},$$

and matrices $\mathbf{A}$, $\mathbf{B}$ and $\mathbf{C}$ by

$$\mathbf{A} = \begin{pmatrix} 11 & -1 & 4 \\ -1 & 11 & 4 \\ 4 & 4 & 14 \end{pmatrix} \quad \mathbf{B} = \begin{pmatrix} 1 & 1 \\ 5 & 2 \\ -2 & 3 \end{pmatrix} \quad \mathbf{C} = \begin{pmatrix} \frac{1}{\sqrt{3}} & \frac{2}{\sqrt{6}} & 0 \\ \frac{1}{\sqrt{3}} & \frac{-1}{\sqrt{6}} & \frac{1}{\sqrt{2}} \\ \frac{1}{\sqrt{3}} & \frac{-1}{\sqrt{6}} & \frac{-1}{\sqrt{2}} \end{pmatrix}.$$

   (a) Find $\mathbf{x}'$, $\mathbf{x}'\mathbf{y}$, $\mathbf{x}\mathbf{y}'$.
   (b) Find $\mathbf{A} + \mathbf{C}$, $\mathbf{A} - \mathbf{C}$, $\mathbf{A}\mathbf{B}$, $|\mathbf{A}|$, $\mathbf{A}\mathbf{1}_3$, $\mathbf{H}\mathbf{A}$.
   (c) Show that $\mathbf{C}$ is orthogonal. Find $\mathbf{C}^{-1}$.
   (d) Find the eigenvalues and eigenvectors of $\mathbf{A}$, and hence the spectral decomposition of $\mathbf{A}$.
   (e) Using the spectral decomposition of $\mathbf{A}$, find $\mathbf{A}^{-1}$ and $\mathbf{A}^{1/2}$. Verify your results by multiplying $\mathbf{A}^{1/2}$ by itself to obtain $\mathbf{A}$, and $\mathbf{A}^{-1}$ by $\mathbf{A}$ to obtain $\mathbf{I}$.
   (f) Find the singular valued decomposition of $\mathbf{B}$.

2. Let

$$\mathbf{A} = \begin{pmatrix} 1 & \rho \\ \rho & 1 \end{pmatrix}.$$

   Find the eigenvalues and eigenvectors of $\mathbf{A}$, and hence the spectral decomposition of $\mathbf{A}$.

3. Show that

$$\frac{\partial \mathbf{x}'\mathbf{x}}{\partial \mathbf{x}} = 2\mathbf{x}, \quad \text{and} \quad \frac{\partial \mathbf{x}'\mathbf{A}\mathbf{x}}{\partial \mathbf{x}} = 2\mathbf{A}\mathbf{x} \quad \text{if } \mathbf{A} \text{ is symmetric.}$$

4. Maximize $6x + 8y$, subject to the constraint $x^2 + y^2 = 1$.

3

Basic multivariate statistics

In this chapter, we start with the basics of multivariate data analysis, and essentially mirror an initial study of univariate statistics, but in the multivariate setting. A study of univariate statistics would begin with the graphical presentation of data, move on to the summary statistics of sample mean, sample median, sample variance, etc., then define various distributions, and finally introduce some statistical inference, such as the use of confidence intervals and t-tests. This is the path taken in this chapter for our multivariate data, except that we leave the graphical illustration of data until the next chapter.

3.1 Summary statistics

We start by looking at the counterpart of the univariate sample mean and other univariate statistics. Let our data be in the form of an $n \times p$ data matrix $\mathbf{X}$, where x_{ri} is the value of the ith variable, for the rth observation. Also, let the rth row of $\mathbf{X}$ be $\mathbf{x}'_r$ (i.e. the rth observation). First, we establish the usual univariate statistics. Let the sample mean for the ith variable be

$$\bar{x}_i = \frac{1}{n} \sum_{r=1}^{n} x_{ri} \qquad (i = 1, \ldots, p).$$

Let the sample variance for the ith variable be s_i^2, but where often we will write this as s_{ii}, and so

$$s_i^2 = s_{ii} = \frac{1}{n-1} \sum_{r=1}^{n} (x_{ri} - \bar{x}_i)^2.$$

Let the sample standard deviation for the ith variable be $s_i = \sqrt{s_{ii}}$. Let s_{ij} be the sample covariance between the ith and jth variables,

$$s_{ij} = \frac{1}{n-1} \sum_{r=1}^{n} (x_{ri} - \bar{x}_i)(x_{rj} - \bar{x}_j).$$

Let r_{ij} be the sample correlation coefficient between the ith and jth variables, and thus

$$r_{ij} = s_{ij}/(s_i s_j).$$

Note that in practice, the word *sample* is often dropped from the terms sample mean, sample variance, etc., but it is always implied so as to distinguish the sample statistics from their population counterparts. We now give these statistics a multivariate role.

The *sample mean vector*, $\bar{\mathbf{x}}$, is defined as

$$\bar{\mathbf{x}} = \begin{pmatrix} \bar{x}_1 \\ \bar{x}_2 \\ \vdots \\ \bar{x}_p \end{pmatrix}.$$

The univariate sample means have simply been placed into a vector. Similarly, the univariate sample variances and covariances are placed into a matrix, $\mathbf{S}$, which is given various names. These are *sample variance matrix*, *sample covariance matrix*, *sample dispersion matrix*, or sometimes *sample variance–covariance matrix*. We will use these names interchangeably. Thus

$$\mathbf{S} = [s_{ij}] = \begin{pmatrix} s_{11} & s_{12} & \cdots & s_{1p} \\ s_{21} & s_{22} & \cdots & s_{2p} \\ \vdots & \vdots & \ddots & \vdots \\ s_{p1} & s_{p2} & \cdots & s_{pp} \end{pmatrix}.$$

The matrix $\mathbf{S}$ is symmetric, as $\text{cov}(X_i, X_j) = \text{cov}(X_j, X_i)$.

Example

Throughout this chapter, we use a dataset on blood pressure from Hand et al. (1994). The data are found at various websites (http://www.stat.ucla.edu/data/ for instance). The data are the systolic and diastolic blood pressure (in mm Hg) for fifteen patients with moderate essential hypertension, pre and post taking the drug captopril. It is a small dataset with only four variables and is ideal for illustrating the statistical concepts as they are discussed.

Let the variables be labelled: x_1 – systolic pre captopril, x_2 – diastolic pre captopril, x_3 – systolic post captopril, and x_4 – diastolic post captopril.

The sample mean vector and sample covariance matrix are

$$\bar{\mathbf{x}} = \begin{pmatrix} 176.9 \\ 112.3 \\ 158.0 \\ 103.1 \end{pmatrix} \quad \mathbf{S} = \begin{pmatrix} 422.9 & 143.2 & 370.8 & 105.1 \\ 143.2 & 109.7 & 153.1 & 96.5 \\ 370.8 & 153.1 & 400.1 & 166.3 \\ 105.1 & 96.5 & 166.3 & 157.6 \end{pmatrix}.$$

We can define $\bar{\mathbf{x}}$ and $\mathbf{S}$ in terms of vectors and matrices, which is useful in calculations within matrix algebra computer packages, or some statistical packages such as MINITAB, S-Plus or SAS. We have

$$\bar{\mathbf{x}} = \frac{1}{n} \sum_{r=1}^{n} \mathbf{x}_r = n^{-1} \mathbf{X}' \mathbf{1}_n.$$

Now

$$s_{ij} = \frac{1}{n-1}\left(\sum_{r=1}^{n} x_{ri}x_{rj} - n\bar{x}_i\bar{x}_j\right),$$

and so

$$S = \frac{1}{n-1}\sum_{r=1}^{n}(\mathbf{x}_r - \bar{\mathbf{x}})(\mathbf{x}_r - \bar{\mathbf{x}})'$$

$$= \frac{1}{n-1}\left(\sum_{r=1}^{n}\mathbf{x}_r\mathbf{x}_r' - n\bar{\mathbf{x}}\bar{\mathbf{x}}'\right)$$

$$= \frac{1}{n-1}\left(\mathbf{X}'\mathbf{X} - n\bar{\mathbf{x}}\bar{\mathbf{x}}'\right)$$

$$= \frac{1}{n-1}(\mathbf{X}'\mathbf{X} - n^{-1}\mathbf{X}'\mathbf{1}_n\mathbf{1}_n'\mathbf{X}),$$

and hence making use of the centring matrix $\mathbf{H}$,

$$S = \frac{1}{n-1}\mathbf{X}'\mathbf{H}\mathbf{X}.$$

The sample correlation matrix, $\mathbf{R}$, is

$$\mathbf{R} = [r_{ij}],$$

and if $\mathbf{D} = \mathrm{diag}(s_i)$, then

$$\mathbf{R} = \mathbf{D}^{-1}\mathbf{S}\mathbf{D}^{-1} \quad \text{or} \quad \mathbf{S} = \mathbf{D}\mathbf{R}\mathbf{D}.$$

The sample covariance matrix, $\mathrm{cov}(\mathbf{x}, \mathbf{y})$, between two random vectors, $\mathbf{x}$ and $\mathbf{y}$, is given by the matrix whose elements are

$$s_{ij} = \frac{1}{n-1}\left(\sum_{r=1}^{n} x_{ri}y_{rj} - n\bar{x}_i\bar{y}_j\right).$$

For example, for the blood pressure data, let $\mathbf{z}_1 = (X_1, X_2)'$ and $\mathbf{z}_2 = (X_3, X_4)'$, and then

$$\mathrm{cov}(\mathbf{z}_1, \mathbf{z}_2) = \begin{pmatrix} 370.8 & 105.1 \\ 153.1 & 96.5 \end{pmatrix}.$$

3.2 Population moments

We now introduce the population moments corresponding to the above sample moments. Let $\mathbf{x}$ be our random vector,

$$\mathbf{x} = \begin{pmatrix} X_1 \\ X_2 \\ \vdots \\ X_p \end{pmatrix}.$$

The population mean vector, $\boldsymbol{\mu}$, is defined as

$$\boldsymbol{\mu} = E(\mathbf{x}) = \begin{pmatrix} E(X_1) \\ E(X_2) \\ \vdots \\ E(X_p) \end{pmatrix} = \begin{pmatrix} \mu_1 \\ \mu_2 \\ \vdots \\ \mu_p \end{pmatrix},$$

where μ_i is the population mean of the univariate random variable X_i.

The population covariance matrix, $\boldsymbol{\Sigma}$ (other names are population variance matrix or population dispersion matrix), is defined as

$$\boldsymbol{\Sigma} = \text{var}(\mathbf{x}) = E[(\mathbf{x} - \boldsymbol{\mu})(\mathbf{x} - \boldsymbol{\mu})'].$$

Hence

$$\boldsymbol{\Sigma} = \begin{pmatrix} E(X_1 - \mu_1)^2 & E(X_1 - \mu_1)(X_2 - \mu_2) & \cdots & E(X_1 - \mu_1)(X_p - \mu_p) \\ E(X_2 - \mu_2)(X_1 - \mu_1) & E(X_2 - \mu_2)^2 & \cdots & E(X_2 - \mu_2)(X_p - \mu_p) \\ \vdots & \vdots & \ddots & \vdots \\ E(X_p - \mu_p)(X_1 - \mu_1) & E(X_p - \mu_p)(X_2 - \mu_2) & \cdots & E(X_p - \mu_p)^2 \end{pmatrix}.$$

Thus

$$\boldsymbol{\Sigma} = [\sigma_{ij}],$$

where $\sigma_{ii} = \text{var}(X_i)$ and $\sigma_{ij} = \text{cov}(X_i, X_j)$. The population correlation matrix is given by

$$\mathbf{P} = [\sigma_{ij}/\sigma_i\sigma_j],$$

where $\sigma_i = \sqrt{\sigma_{ii}}$, the population standard deviation of X_i. Again, the word population is often dropped from the terms for these population moments, and we say 'mean', or 'dispersion matrix', the term 'population' being implied.

The covariance matrix between two random vectors, $\mathbf{X}$ and $\mathbf{Y}$, is similarly defined as

$$\text{cov}(\mathbf{x}, \mathbf{y}) = E[(\mathbf{x} - \boldsymbol{\mu_x})(\mathbf{y} - \boldsymbol{\mu_y})'].$$

Note that if the random vectors $\mathbf{x}$ and $\mathbf{y}$ were combined into one vector, $\mathbf{z}$ say, then $\text{cov}(\mathbf{x}, \mathbf{y})$ would simply be a submatrix of the variance matrix of $\mathbf{z}$.

3.2.1 Linear combinations

A linear combination, based on random vector $\mathbf{x}$, is a weighted sum of the elements of $\mathbf{x}$. Consider the linear combination $Y = \mathbf{a}'\mathbf{x} = a_1X_1 + a_2X_2 + \ldots + a_pX_p$, where $\mathbf{a}$ is a vector of constants. Then

$$\begin{aligned} E(Y) &= E(a_1X_1 + a_2X_2 + \ldots + a_pX_p) \\ &= a_1E(X_1) + a_2E(X_2) + \ldots + a_pE(X_p) \\ &= a_1\mu_1 + a_2\mu_2 + \ldots + a_p\mu_p, \end{aligned}$$

and therefore

$$E(Y) = \mathbf{a}'\boldsymbol{\mu},$$

where $\boldsymbol{\mu}$ is the mean vector of $\mathbf{x}$.

The variance of Y is given by

$$\begin{aligned}
\text{var}(Y) &= E[(Y - E(Y))^2] \\
&= E[(\mathbf{a}'\mathbf{x} - \mathbf{a}'\boldsymbol{\mu})^2] \\
&= E[(\mathbf{a}'(\mathbf{x} - \boldsymbol{\mu})(\mathbf{x} - \boldsymbol{\mu})'\mathbf{a}] \\
&= E\left[\mathbf{a}' \begin{pmatrix} (X_1 - \mu_1)^2 & \cdots & (X_1 - \mu_1)(X_p - \mu_p) \\ \vdots & \ddots & \vdots \\ (X_p - \mu_p)(X_1 - \mu_1) & \cdots & (X_p - \mu_p)^2 \end{pmatrix} \mathbf{a} \right].
\end{aligned}$$

Thus

$$\text{var}(Y) = \mathbf{a}'\boldsymbol{\Sigma}\mathbf{a},$$

where $\boldsymbol{\Sigma}$ is the covariance matrix of $\mathbf{x}$.

Now consider q linear combinations of $\mathbf{x}$,

$$Y_1 = \mathbf{a}_1'\mathbf{x}, \ Y_2 = \mathbf{a}_2'\mathbf{x}, \ldots, \ Y_q = \mathbf{a}_q'\mathbf{x}.$$

Place these Y's into a vector, so that $\mathbf{y} = (Y_1, Y_2, \ldots, Y_q)'$, and place the constant vectors into a matrix, so that $\mathbf{A} = (\mathbf{a}_1, \mathbf{a}_2, \ldots, \mathbf{a}_q)$. The matrix $\mathbf{A}$ is called a transformation matrix. Then

$$\mathbf{y} = \mathbf{A}'\mathbf{x},$$

and then

$$E(\mathbf{y}) = \mathbf{A}'\boldsymbol{\mu}, \quad \text{var}(\mathbf{y}) = \mathbf{A}'\boldsymbol{\Sigma}\mathbf{A}.$$

Example

Let $\mathbf{x}$ have mean vector and covariance matrix given by

$$\boldsymbol{\mu} = \begin{pmatrix} 1 \\ 2 \\ 3 \end{pmatrix}, \quad \boldsymbol{\Sigma} = \begin{pmatrix} 4 & 0 & 0 \\ 0 & 4 & 0 \\ 0 & 0 & 1 \end{pmatrix}.$$

If $Y_1 = X_1 + X_2 + X_3$ and $Y_2 = 2X_1 - X_2 - X_3$, the transformation matrix is

$$\mathbf{A} = \begin{pmatrix} 1 & 2 \\ 1 & -1 \\ 1 & -1 \end{pmatrix}.$$

Then

$$E(\mathbf{y}) = \mathbf{A}'\boldsymbol{\mu} = \begin{pmatrix} 1 & 1 & 1 \\ 2 & -1 & -1 \end{pmatrix} \begin{pmatrix} 1 \\ 2 \\ 3 \end{pmatrix} = \begin{pmatrix} 6 \\ -3 \end{pmatrix},$$

and

$$\text{var}(\mathbf{y}) = \begin{pmatrix} 1 & 1 & 1 \\ 2 & -1 & -1 \end{pmatrix} \begin{pmatrix} 4 & 0 & 0 \\ 0 & 4 & 0 \\ 0 & 0 & 1 \end{pmatrix} \begin{pmatrix} 1 & 2 \\ 1 & -1 \\ 1 & -1 \end{pmatrix} = \begin{pmatrix} 9 & 3 \\ 3 & 21 \end{pmatrix}.$$

The above transformation results for population mean vectors and population covariance matrices also hold for sample mean vectors and sample covariance matrices.

3.3 Multivariate distributions

In this section we look at two multivariate distributions, the multinomial and the multivariate normal distribution. One problem with modelling multivariate data with a probability distribution is that often the variables upon which we collect the data are very disparate in their nature. For instance, suppose the variables of interest are ethnic group, weight, number of children, and time to recovery from a certain illness. This is a rather contrived example, but it would be very difficult to model these four variables together probabilistically. Ethnic group is a categorical variable with a probability associated with each group, weight would be normally distributed (or a mixture of normals because of gender), number of children might be modelled by a Poisson distribution, and time to recovery, by a gamma distribution. The problem is to bring the variables together, and not only model the marginal distributions, but also the correlation and other interactions between the variables. Because of this difficulty, most multivariate distributions are based on variables that are of the same nature, for instance all the variables having a univariate normal distribution.

3.3.1 Multinomial distribution

The multinomial distribution is the multivariate generalization of the binomial distribution. Think of a misshapen p-sided die, where unlike an ordinary die, the outcomes (scores), when thrown, are not equiprobable. Let the probability that it lands with face i uppermost, and hence gives a score of i, equal π_i ($\sum_i \pi_i = 1$). Suppose the die is rolled N times, and then the probability of obtaining x_i scores of i ($i = 1, \ldots, p$) is

$$\frac{N!}{x_1! x_2! \ldots x_p!} \pi_1^{x_1} \pi_2^{x_2} \ldots \pi_p^{x_p} \qquad \left(\sum_{i=1}^{p} x_i = N; \; \sum_{i=1}^{p} \pi_i = 1 \right).$$

For $p = 2$, this becomes the probability function for the binomial distribution. Of course, using a die in this way is only to introduce the distribution. The distribution is appropriate for any situation where observations fall into one of p categories, and where observations are independent. The x_i's can be placed in a vector $\mathbf{x}$ to emphasize the multivariate nature of the distribution. The mean vector and covariance matrix can be shown to be

$$\boldsymbol{\mu} = \begin{pmatrix} N\pi_1 \\ N\pi_2 \\ \vdots \\ N\pi_p \end{pmatrix}, \quad \boldsymbol{\Sigma} = \begin{pmatrix} N\pi_1(1-\pi_1) & -N\pi_1\pi_2 & \cdots & -N\pi_1\pi_p \\ -N\pi_2\pi_1 & N\pi_2(1-\pi_2) & \cdots & -N\pi_2\pi_p \\ \vdots & \vdots & \ddots & \vdots \\ -N\pi_p\pi_1 & -N\pi_p\pi_2 & \cdots & N\pi_p(1-\pi_p) \end{pmatrix}.$$

Example

Each day, one hundred cars are produced on a production line. Each is classified as one of P_0, P_1, P_2, or F, where P_j is *pass* with j minor faults, and F is *fail*. Let X_1 be the number of cars that pass with 0 faults, X_2 the number that pass with 1 fault, X_3 the number that pass with 2 faults, and X_4 the number that fail. Suppose $\pi_1 = 0.3$, $\pi_2 = 0.5$, $\pi_3 = 0.15$, and $\pi_4 = 0.05$. Then

$$\mu = \begin{pmatrix} 30 \\ 50 \\ 15 \\ 5 \end{pmatrix}, \quad \Sigma = \begin{pmatrix} 9 & -15 & -4.5 & -1.5 \\ -15 & 25 & -7.5 & -2.5 \\ -4.5 & -7.5 & 2.25 & -0.75 \\ -1.5 & -2.5 & -0.75 & 0.25 \end{pmatrix}.$$

In practice, the π_i's generally will not be known and are estimated from a sample by $\hat{\pi}_j = n^{-1} \sum_{r=1}^{n} n_{rj}/N$, where n is the sample size, and n_{rj} is the number in class j in the rth sample.

Krzanowski and Marriott (1994) and Johnson and Kotz (1969, 1972) give more multivariate generalizations of well-known univariate discrete distributions as well as for continuous distributions.

3.3.2 Multivariate normal distribution

The multivariate normal distribution plays a similar role in multivariate statistics as does the univariate normal distribution in univariate statistics. A random vector, $\mathbf{x}$, has a multivariate normal distribution (MVN), if its probability density function (pdf) is given by

$$f(\mathbf{x}) = \frac{1}{(2\pi)^{p/2}|\Sigma|^{1/2}} \exp\left\{-\tfrac{1}{2}(\mathbf{x} - \mu)'\Sigma^{-1}(\mathbf{x} - \mu)\right\},$$

where μ is a vector of constants, $(\mu_1, \mu_2, \ldots, \mu_p)'$, and Σ is a $p \times p$ symmetric positive definite matrix of constants

$$\Sigma = \begin{pmatrix} \sigma_{11} & \sigma_{12} & \cdots & \sigma_{1p} \\ \sigma_{21} & \sigma_{22} & \cdots & \sigma_{2p} \\ \vdots & \vdots & \ddots & \vdots \\ \sigma_{p1} & \sigma_{p2} & \cdots & \sigma_{pp} \end{pmatrix}.$$

One can guess that the constant vector and matrix have been labelled as μ and Σ as they will turn out to be the population mean vector and covariance matrix – but that comes later.

For $p = 2$ (the bivariate normal), the expansion of $f(\mathbf{x})$ is

$$f(x_1, x_2) = \frac{1}{2\psi_1\sigma_2\sqrt{1-\rho^2}}$$

$$\times \exp\left\{-\frac{1}{2(1-\rho^2)}\left[\frac{(x_1-\mu_1)^2}{\sigma_1^2}\right.\right.$$

$$-\frac{2\rho(x_1-\mu_1)(x_2-\mu_2)}{\sigma_1\sigma_2}$$

$$+\left.\left.\frac{(x_2-\mu_2)^2}{\sigma_2^2}\right]\right\},$$

where $\rho = \sigma_{12}/\sigma_1\sigma_2$. This expanded form shows the similarity with the univariate normal distribution. For $\rho = 0$, the pdf becomes the product of two univariate normal distributions, showing that zero correlation implies independence. This is also true for any p, that, if all correlations are zero (i.e. all $\sigma_{ij} = 0, i \neq j$), then the X_i's are independent. This is only true for the multivariate normal distribution as zero correlation does not imply independence generally.

If $\mathbf{x}$ has a multivariate normal distribution, we write $\mathbf{x} \sim N_p(\boldsymbol{\mu}, \boldsymbol{\Sigma})$. The following are some properties of the distribution.

(i) Contours of equal probability. The contours of equal probability are the curves defined by $f(\mathbf{x}) = k$, where k is a constant. Equating $f(\mathbf{x})$ to k, we have

$$\frac{1}{(2\pi)^{p/2}|\boldsymbol{\Sigma}|^{1/2}}\exp\left\{-\tfrac{1}{2}(\mathbf{x}-\boldsymbol{\mu})'\boldsymbol{\Sigma}^{-1}(\mathbf{x}-\boldsymbol{\mu})\right\} = k.$$

Simplifying this, we have

$$(\mathbf{x}-\boldsymbol{\mu})'\boldsymbol{\Sigma}^{-1}(\mathbf{x}-\boldsymbol{\mu}) = c, \tag{3.1}$$

where c is a constant. Let $\mathbf{y} = \mathbf{x} - \boldsymbol{\mu}$, and then equation (3.1) becomes

$$\mathbf{y}'\boldsymbol{\Sigma}^{-1}\mathbf{y} = c.$$

Put $\boldsymbol{\Sigma}$ in terms of its spectral decomposition, $\boldsymbol{\Sigma} = \mathbf{V}\boldsymbol{\Lambda}\mathbf{V}'$, and thus

$$\mathbf{y}'\mathbf{V}\boldsymbol{\Lambda}^{-1}\mathbf{V}'\mathbf{y} = c.$$

Now let $\mathbf{z} = \mathbf{V}'\mathbf{y}$, and so

$$\mathbf{z}'\boldsymbol{\Lambda}^{-1}\mathbf{z} = c. \tag{3.2}$$

Now $\boldsymbol{\Lambda}^{-1} = \text{diag}(\lambda_1^{-1}, \lambda_2^{-1}, \ldots, \lambda_p^{-1})$, and hence equation (3.2) becomes

$$z_1^2\frac{1}{\lambda_1} + z_2^2\frac{1}{\lambda_2} + \ldots + z_p^2\frac{1}{\lambda_p} = c.$$

Thus we have transformed to principal axes, and the contours of equal probability are ellipsoids. Figure 3.1 shows a plot of the density function of a bivariate normal distribution with

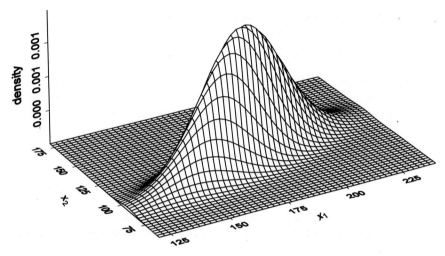

Figure 3.1 Bivariate normal density

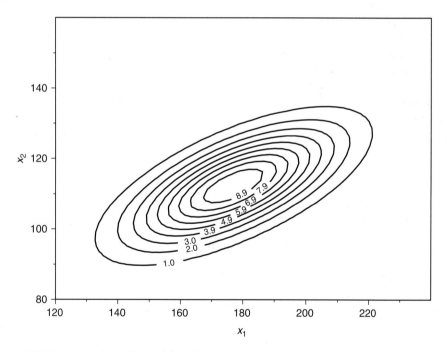

Figure 3.2 Contours of equal probability (density × 1000)

mean vector $(176.9, 112.3)'$, and covariance matrix with elements $\sigma_{11} = 400.1$, $\sigma_{22} = 157.6$, and $\sigma_{12} = 166.3$. Figure 3.2 shows the contours of equal probability. These figures were drawn using S-Plus. The density diagram is typical for a bivariate normal distribution, with the maximum placed at the population mean, and the shape determined by the population variances and the correlation. If X_1 and X_2 were independent (or simply uncorrelated), then the contour ellipses would have their axes parallel to the x_1 axis and x_2 axis. If the variances were also equal, then the contours of equal probability would be circles. The value of the correlation coefficient determines the slope of the long axes of the contour ellipses. Drawing the pdf for a $\mathbf{x}$ in higher dimension than two is an impossible task.

(ii) The distribution of a linear transformation. First we find the distribution of $\boldsymbol{\Sigma}^{-1/2}(\mathbf{x} - \boldsymbol{\mu})$, which will help in establishing several properties of the multivariate normal distribution. Make the transformation $\mathbf{y} = \mathbf{x} - \boldsymbol{\mu}$, whereupon the pdf for $\mathbf{Y}$ is

$$f(\mathbf{y}) = \frac{1}{(2\pi)^{p/2}|\boldsymbol{\Sigma}|^{1/2}} \exp\left\{-\tfrac{1}{2}\mathbf{y}'\boldsymbol{\Sigma}^{-1}\mathbf{y}\right\}.$$

Note that we will use $f(.)$ for all pdf's, not distinguishing between different ones by using different letters.

We now make the transformation $\mathbf{z} = \boldsymbol{\Sigma}^{-1/2}\mathbf{y}$, recalling that $\boldsymbol{\Sigma}^{-1/2} = \mathbf{V}\boldsymbol{\Lambda}^{-1/2}\mathbf{V}'$, based on the spectral decomposition of $\boldsymbol{\Sigma}$. The Jacobian for the transformation is $|\boldsymbol{\Sigma}^{1/2}|$, and hence the pdf for $\mathbf{z}$ is

$$f(\mathbf{z}) = |\boldsymbol{\Sigma}|^{1/2} \frac{1}{(2\pi)^{p/2}|\boldsymbol{\Sigma}|^{1/2}} \exp\left\{-\tfrac{1}{2}(\boldsymbol{\Sigma}^{1/2}\mathbf{z})'\boldsymbol{\Sigma}^{-1}(\boldsymbol{\Sigma}^{1/2}\mathbf{z})\right\}$$

$$= \frac{1}{(2\pi)^{p/2}} \exp\left\{-\tfrac{1}{2}\mathbf{z}'\mathbf{z}\right\}$$

$$= \prod_{i=1}^{p} \frac{1}{(2\pi)^{1/2}} \exp\left(-\tfrac{1}{2}z_i^2\right).$$

Hence the Z_i's are independent and each has a $N(0, 1)$ distribution. Thus, we have shown that

$$\boldsymbol{\Sigma}^{-1/2}(\mathbf{x} - \boldsymbol{\mu}) \sim N_p(\mathbf{0}, \mathbf{I}).$$

It is also possible to show that the distribution of $\mathbf{y} = \mathbf{B}\mathbf{x} + \mathbf{b}$, where $\mathbf{B}$ is a $q \times p$ matrix of constants, and $\mathbf{b}$ is a constant vector of length q, is also multivariate normal, with

$$\mathbf{y} = \mathbf{B}\mathbf{x} + \mathbf{b} \sim N_q(\mathbf{B}\boldsymbol{\mu} + \mathbf{b}, \mathbf{B}\boldsymbol{\Sigma}\mathbf{B}').$$

We will not prove this here; the simplest most elegant proof is via characteristic functions.

(iii) Mean vector and covariance matrix. The ith element of the mean vector for $\mathbf{x}$ is given by

$$\int_{-\infty}^{\infty} \cdots \int_{-\infty}^{\infty} x_i f(\mathbf{x}) \, d\mathbf{x},$$

which can be evaluated using transformations similar to those in the previous section. It is no surprise that $E(\mathbf{x}) = \boldsymbol{\mu}$.

Similarly, the value of $E[(X_i - \mu_i)(X_j - \mu_j)]$ is found to be σ_{ij}, and hence $\text{var}(\mathbf{x}) = \boldsymbol{\Sigma}$.

(iv) Marginal distributions. The marginal distribution of X_i is $N(\mu_i, \sigma_{ii})$. This is easily seen by choosing $\mathbf{B}$ in the transformation in the above section to be a matrix of zeros, except for $b_{ii} = 1$.

Now suppose $\mathbf{x}$ is partitioned as follows

$$\mathbf{x} = \begin{pmatrix} \mathbf{x}_1 \\ \mathbf{x}_2 \end{pmatrix},$$

where $\mathbf{x}_1$ has q elements, and $\mathbf{x}_2$, $p - q$ elements. Also partition the mean vector and covariance matrix appropriately as

$$\boldsymbol{\mu} = \begin{pmatrix} \boldsymbol{\mu}_1 \\ \boldsymbol{\mu}_2 \end{pmatrix} \qquad \boldsymbol{\Sigma} = \begin{pmatrix} \boldsymbol{\Sigma}_{11} & \boldsymbol{\Sigma}_{12} \\ \boldsymbol{\Sigma}_{21} & \boldsymbol{\Sigma}_{22} \end{pmatrix}.$$

Then by choosing $\mathbf{B}$ to have zeros everywhere, except for the first q elements of the diagonal, which are given the value unity, it is easily seen that the marginal distribution of $\mathbf{x}_1$ is $N(\boldsymbol{\mu}_1, \boldsymbol{\Sigma}_{11})$.

(v) Conditional distributions. The conditional distribution of $\mathbf{x}_1$, given $\mathbf{x}_2$, is multivariate normal with

$$E(\mathbf{x}_1 | \mathbf{x}_2) = \boldsymbol{\mu}_1 + \boldsymbol{\Sigma}_{12} \boldsymbol{\Sigma}_{22}^{-1} (\mathbf{x}_2 - \boldsymbol{\mu}_2)$$

$$\mathrm{var}(\mathbf{x}_1 | \mathbf{x}_2) = \boldsymbol{\Sigma}_{11} - \boldsymbol{\Sigma}_{12} \boldsymbol{\Sigma}_{22}^{-1} \boldsymbol{\Sigma}_{21}.$$

We do not give a proof of this result. Note that the matrix $\boldsymbol{\Sigma}_{22}$ must not be singular.

3.3.3 The distribution of a sum of $N_p(\boldsymbol{\mu}, \boldsymbol{\Sigma})$ rv's

If $\mathbf{y} = \sum_{i=1}^{n} a_i \mathbf{x}_i$ is a weighted sum of p-dimensional multivariate normal random vectors, $\mathbf{x}_i$, the a_i's being scalars, then $\mathbf{y}$ also has a multivariate normal distribution, with

$$E(\mathbf{y}) = \sum_{i=1}^{n} a_i E(\mathbf{x}_i)$$

$$\mathrm{var}(\mathbf{y}) = \sum_{i=1}^{n} a_i^2 \mathrm{var}(\mathbf{x}_i) + 2 \sum_{i<j}^{n} a_i a_j \mathrm{cov}(\mathbf{x}_i, \mathbf{x}_j).$$

Again, the easiest proof is via characteristic functions. When the $\mathbf{x}_i$ are independent (or simply uncorrelated), then the term involving $\mathrm{cov}(\mathbf{x}_i, \mathbf{x}_j)$ vanishes.

3.3.4 The Wishart distribution

The Wishart distribution is the multivariate generalization of the χ^2 distribution. Let $\mathbf{x}_i$ $(i = 1, \ldots, n)$ be independently distributed as $N_p(\mathbf{0}, \boldsymbol{\Sigma})$ distributions. Then $\mathbf{W} = \sum_{i=1}^{n} \mathbf{x}_i \mathbf{x}_i'$ has a Wishart distribution with n degrees of freedom. The matrix $\mathbf{W}$ is called the Wishart matrix and we write $\mathbf{W} \sim W_p(n, \boldsymbol{\Sigma})$. The pdf of the Wishart distribution is

$$\frac{|\mathbf{W}|^{(n-p-1)/2} \exp(-\frac{1}{2} \mathrm{tr} \mathbf{W} \boldsymbol{\Sigma}^{-1})}{2^{np/2} \pi^{1/4} |\boldsymbol{\Sigma}|^{n/2} \prod_{i=1}^{p} \Gamma[(\frac{1}{2}(n + 1 - i)]}.$$

This compares to the univariate result that if $Y_1, \ldots, Y_n \sim N(0, \sigma^2)$, then $\sum Y_i^2 \sim \sigma^2 \chi_n^2$.

If $\mathbf{W}_1$ and $\mathbf{W}_2$ are independent, with $\mathbf{W}_1 \sim W_p(n_1, \boldsymbol{\Sigma})$ and $\mathbf{W}_2 \sim W_p(n_2, \boldsymbol{\Sigma})$, then $\mathbf{W}_1 + \mathbf{W}_2 \sim W_p(n_1 + n_2, \boldsymbol{\Sigma})$.

3.3.5 Hotelling's T^2-distribution

Hotelling's T^2-distribution is the multivariate generalization of the Student's t-distribution. Let $\mathbf{x}$ and $\mathbf{W}$ be independent, with $\mathbf{x} \sim N_p(\boldsymbol{\mu}, \boldsymbol{\Sigma})$, and $\mathbf{W} \sim W_p(n, \boldsymbol{\Sigma})$, with $n > p - 1$. Let $T_p^2(n) = n(\mathbf{x} - \boldsymbol{\mu})'\mathbf{W}^{-1}(\mathbf{x} - \boldsymbol{\mu})$. Then $T_p^2(n)$ is said to have an Hotelling's T^2-distribution, with parameters n and p.

Hotelling's T^2-distribution is closely connected to the F-distribution $(T_p^2(n) = np$ $(n - p + 1)^{-1}F_{p,n-p+1})$. So in order to calculate percentage points of the $T_p^2(n)$ distribution, we can use the F-tables, with

$$P\{T_p^2(n) \geq T_\alpha\} = P\{F_{p,n-p+1} \geq (np)^{-1}(n - p + 1)T_\alpha\}.$$

3.3.6 The distribution of $\bar{\mathbf{x}}$ and S

We are now in a position to find the distribution of $\bar{\mathbf{x}}$ and $\mathbf{S}$ for a random sample drawn from a multivariate normal distribution. Since $\bar{\mathbf{x}} = \frac{1}{n}\sum_{i=1}^{n}\mathbf{x}_i$ is a sum of the random vectors $\mathbf{x}_i$, them from (3.3.3) above, $\bar{\mathbf{x}} \sim N_p(\boldsymbol{\mu}, n^{-1}\boldsymbol{\Sigma})$.

We now consider the distribution of $(\mathbf{x} - \boldsymbol{\mu})'\boldsymbol{\Sigma}^{-1}(\mathbf{x} - \boldsymbol{\mu})$. From (ii) above, $\mathbf{y} = \boldsymbol{\Sigma}^{-1/2}(\mathbf{x} - \boldsymbol{\mu}) \sim N_p(\mathbf{0}, \mathbf{I})$, and hence

$$(\mathbf{x} - \boldsymbol{\mu})'\boldsymbol{\Sigma}^{-1}(\mathbf{x} - \boldsymbol{\mu}) = \mathbf{y}'\mathbf{y} = \sum_{i=1}^{p} Y_i^2 \sim \chi_p^2,$$

since Y_i^2 is the square of a N(0, 1) random variable, and thus has a χ_1^2 distribution, and the sum of p independent χ_1^2 random variables has a χ_p^2 distribution.

Now we find the distribution of $\mathbf{S}$ and show it is independent of the distribution of $\bar{\mathbf{x}}$.

We write our data matrix as

$$\mathbf{X} = \begin{pmatrix} \mathbf{x}_1' \\ \mathbf{x}_2' \\ \vdots \\ \mathbf{x}_n' \end{pmatrix},$$

and introduce the transformation $\mathbf{Y} = (\mathbf{y}_1\ \mathbf{y}_2 \ldots \mathbf{y}_n)' = \mathbf{B}'\mathbf{X}$, where

$$\mathbf{B} = \begin{pmatrix} \frac{1}{\sqrt{n}} & \frac{1}{\sqrt{2\times1}} & \frac{1}{\sqrt{3\times2}} & \frac{1}{\sqrt{4\times3}} & \cdots & \frac{1}{\sqrt{n\times(n-1)}} \\ \frac{1}{\sqrt{n}} & \frac{-1}{\sqrt{2\times1}} & \frac{1}{\sqrt{3\times2}} & \frac{1}{\sqrt{4\times3}} & \cdots & \frac{1}{\sqrt{n\times(n-1)}} \\ \frac{1}{\sqrt{n}} & 0 & \frac{-2}{\sqrt{3\times2}} & \frac{1}{\sqrt{4\times3}} & \cdots & \frac{1}{\sqrt{n\times(n-1)}} \\ \frac{1}{\sqrt{n}} & 0 & 0 & \frac{-3}{\sqrt{4\times3}} & \cdots & \frac{1}{\sqrt{n\times(n-1)}} \\ \vdots & \vdots & \vdots & \vdots & \ddots & \vdots \\ \frac{1}{\sqrt{n}} & 0 & 0 & 0 & \cdots & \frac{-(n-1)}{\sqrt{n\times(n-1)}} \end{pmatrix}.$$

Matrix $\mathbf{B}$ is called a Helmert matrix. It is an orthogonal matrix, and so $\mathbf{B}' = \mathbf{B}^{-1}$ and $\mathbf{BB}' = \mathbf{B}'\mathbf{B} = \mathbf{I}$. The first row of $\mathbf{Y}$ gives $\mathbf{y}_1 = \frac{1}{\sqrt{n}}\sum_{i=1}^{n}\mathbf{x}_i = \sqrt{n}\,\bar{\mathbf{x}}$, and for the other rows

$$\mathbf{y}_j = \frac{1}{\sqrt{j(j-1)}}\sum_{i=1}^{j-1}\mathbf{x}_i - \frac{(j-1)}{\sqrt{j(j-1)}}\mathbf{x}_j.$$

Now the $\mathbf{y}$'s are independent since $\mathbf{B}$ is orthogonal. As a simple check on this, we calculate $\text{cov}(\mathbf{y}_2, \mathbf{y}_3)$.

$$\begin{aligned}
\text{cov}(\mathbf{y}_2, \mathbf{y}_3) &= \text{cov}\left(\frac{1}{\sqrt{2}}\mathbf{x}_1 - \frac{1}{\sqrt{2}}\mathbf{x}_2, \frac{1}{\sqrt{3}}\mathbf{x}_1 + \frac{1}{\sqrt{3}}\mathbf{x}_2 - \frac{2}{\sqrt{3}}\mathbf{x}_3\right) \\
&= \frac{1}{\sqrt{6}}\left(\text{var}(\mathbf{x}_1) - \text{var}(\mathbf{x}_2) - 2\text{cov}(\mathbf{x}_1, \mathbf{x}_3) + 2\text{cov}(\mathbf{x}_2, \mathbf{x}_3)\right) \\
&= \mathbf{0},
\end{aligned}$$

since $\text{var}(\mathbf{x}_1) = \text{var}(\mathbf{x}_2) = \boldsymbol{\Sigma}$, and $\text{cov}(\mathbf{x}_1, \mathbf{x}_3) = \text{cov}(\mathbf{x}_2, \mathbf{x}_3) = \mathbf{0}$. It is easily seen that the distribution of $\mathbf{y}_1$ is $N_p(\sqrt{n}\,\boldsymbol{\mu}, \boldsymbol{\Sigma})$, and for the other $\mathbf{y}$'s, we have $\mathbf{y}_j \sim N_p(\mathbf{0}, \boldsymbol{\Sigma})$.

Now consider

$$\mathbf{X}'\mathbf{X} = \mathbf{X}'(\mathbf{I})\mathbf{X} = \mathbf{X}'\mathbf{BB}'\mathbf{X} = \mathbf{Y}'\mathbf{Y} = \sum_{i=1}^{n}\mathbf{y}_i'\mathbf{y}_i = \mathbf{y}_1'\mathbf{y}_1 + \sum_{i=2}^{n}\mathbf{y}_i'\mathbf{y}_i$$

$$= n\bar{\mathbf{x}}'\bar{\mathbf{x}} + \sum_{i=2}^{n}\mathbf{y}_i'\mathbf{y}_i.$$

Hence

$$(n-1)\mathbf{S} = \mathbf{X}'\mathbf{X} - n\bar{\mathbf{x}}'\bar{\mathbf{x}} = \sum_{i=2}^{n}\mathbf{y}_i'\mathbf{y}_i.$$

Now since $\bar{\mathbf{x}}$ is based on $\mathbf{y}_1$, and $\mathbf{S}$ is based on $\mathbf{y}_2, \ldots, \mathbf{y}_n$, $\bar{\mathbf{x}}$ and $\mathbf{S}$ are independent, since the $\mathbf{y}_i$'s are independent. Also, the form of $(n-1)\mathbf{S}$ in terms of the sum of the $\mathbf{y}_i$'s shows that $(n-1)\mathbf{S} \sim W_p(n-1, \boldsymbol{\Sigma})$, and hence the result.

We now consider the distribution of $n(\bar{\mathbf{x}} - \boldsymbol{\mu})'\mathbf{S}^{-1}(\bar{\mathbf{x}} - \boldsymbol{\mu})$. We have

$$\sqrt{n}\,\bar{\mathbf{x}} \sim N_p(\sqrt{n}\,\boldsymbol{\mu}, \boldsymbol{\Sigma})$$
$$(n-1)\mathbf{S} \sim W_p(n-1, \boldsymbol{\Sigma}),$$

and therefore

$$(n-1)(\sqrt{n}\,\bar{\mathbf{x}} - \sqrt{n}\,\boldsymbol{\mu})'\frac{\mathbf{S}^{-1}}{n-1}(\sqrt{n}\,\bar{\mathbf{x}} - \sqrt{n}\,\boldsymbol{\mu}) \sim T_p^2(n-1),$$

and after rearranging, we have the result

$$n(\bar{\mathbf{x}} - \boldsymbol{\mu})'\mathbf{S}^{-1}(\bar{\mathbf{x}} - \boldsymbol{\mu}) \sim T_p^2(n-1).$$

Compare these results with the analogous results in the univariate case, namely, $\bar{X}$ and $(n-1)s^2$ are independently distributed, with

$$\bar{X} \sim N(\mu, n^{-1}\sigma^2)$$

$$(n-1)s^2 \sim \sigma^2 \chi_{n-1}^2$$

$$\frac{\bar{x}}{s/\sqrt{n}} \sim t_{n-1}.$$

3.3.7 Central limit theorem

If $x_1, x_2, \ldots, x_n$ is a random sample from a distribution with mean vector μ and covariance matrix Σ, then $\sqrt{n}(\bar{x} - \mu)$ converges in distribution to $N_p(0, \Sigma)$. In practical terms, the sample mean vector has approximately a $N_p(\mu, n^{-1}\Sigma)$ distribution.

3.4 Statistical inference

We now look at some statistical inference techniques based on the multivariate normal distribution that mirror those in univariate statistics for the univariate normal distribution, but without going deeply into the distributional theory underlying the various methods. We cover the three areas of *point estimation, confidence regions*, and *hypothesis testing*.

3.4.1 Point estimation: maximum likelihood

The likelihood, $L(\mu, \Sigma)$, of μ and Σ, based on a random sample, $x_1, x_2, \ldots, x_n$, from a $N_p(\mu, \Sigma)$ distribution, is the joint distribution of $x_1, x_2, \ldots, x_n$, viewed as a function of μ and Σ. Thus

$$L(\mu, \Sigma) = \prod_{r=1}^{n} \frac{1}{(2\pi)^{p/2}|\Sigma|^{1/2}} \exp\left\{-\tfrac{1}{2}(x_r - \mu)'\Sigma^{-1}(x_r - \mu)\right\}.$$

The log-likelihood function, $l(\mu, \Sigma) = \ln L(\mu, \Sigma)$, is

$$l(\mu, \Sigma) = \tfrac{1}{2}np \ln 2\pi - \tfrac{1}{2}n \ln |\Sigma| - \tfrac{1}{2} \sum_{r=1}^{n} (x_r - \mu)'\Sigma^{-1}(x_r - \mu).$$

The maximum likelihood estimates, $\hat{\mu}$ and $\hat{\Sigma}$, of μ and Σ, are those values of μ and Σ that maximize the likelihood, or equivalently the log-likelihood. Differentiating $l(\mu, \Sigma)$ with respect to μ, and equating to 0, we obtain

$$\frac{\partial l}{\partial \mu} = -\tfrac{1}{2} \sum_{r=1}^{n} \Sigma^{-1}(x_r - \mu) = 0,$$

and hence

$$\hat{\mu} = \bar{x}.$$

It can be shown that the maximum likelihood estimate of Σ is given by

$$\hat{\Sigma} = \frac{(n-1)}{n}\mathbf{S}.$$

The algebra involved in finding this is rather lengthy, and the reader is referred to books such as Mardia et al. (1979). Note that it also has to be shown that a maximum of the log-likelihood has been found, and not a minimum, nor a point of inflection, but again, this is left to other books.

Note that $\hat{\Sigma}$ is a biased estimator of Σ, and we will usually use $\mathbf{S}$, which is an unbiased estimator of Σ.

Example

For the blood pressure data,

$$\hat{\mu} = \bar{\mathbf{x}} = \begin{pmatrix} 176.9 \\ 112.3 \\ 158.0 \\ 103.1 \end{pmatrix}, \quad \hat{\Sigma} = \frac{n-1}{n}\mathbf{S} = \begin{pmatrix} 394.7 & 133.6 & 346.1 & 98.1 \\ 133.6 & 102.4 & 142.9 & 90.1 \\ 346.1 & 142.9 & 373.5 & 155.2 \\ 98.1 & 90.1 & 155.2 & 147.1 \end{pmatrix}.$$

Using $\mathbf{S}$ rather than $\hat{\Sigma}$ to estimate Σ, the estimated approximate distribution for the sample mean vector, $\bar{\mathbf{x}}$, is $N_4(\bar{\mathbf{x}}, \mathbf{S}/15)$.

3.4.2 Confidence regions for μ

The reader will be familiar with the notion of a confidence interval for an unknown parameter, the most commonly seen being the one based on the t-distribution, for the population mean. The $100(1-\alpha)\%$ confidence interval for μ is

$$\left[\bar{x} - t_{n-1,\alpha/2}\, s/\sqrt{n},\ \bar{x} + t_{n-1,\alpha/2}\, s/\sqrt{n} \right],$$

where $t_{k,\alpha}$ is the upper 100α percentage point of the t-distribution, with k degrees of freedom. The confidence interval is established using the pivotal quantity, $\sqrt{n}(\bar{X} - \mu)/s$, which has a t-distribution. The key idea is that the distribution of the pivotal quantity does not involve the parameter, and is then used to find a random interval, and hence the confidence interval.

When there is more than one parameter, we can find a confidence interval for each parameter. However, in general, the random intervals used to obtain the confidence intervals will not be independent. An alternative is to produce a confidence region in a Euclidean space, with dimension equal to the number of parameters. As with confidence intervals, confidence regions are not unique. We consider two cases for the multivariate normal distribution.

(i) Σ known.   When Σ is known, the pivotal quantity is $n(\bar{\mathbf{x}} - \mu)'\Sigma^{-1}(\bar{\mathbf{x}} - \mu)$, which we have seen to have a χ_p^2 distribution. This leads to the $100(1-\alpha)\%$ confidence region for μ,

$$\left\{ \mu : n(\mu - \bar{\mathbf{x}})'\Sigma^{-1}(\mu - \bar{\mathbf{x}}) \leq \chi_{p,\alpha}^2 \right\},$$

where $\chi^2_{p,\alpha}$ is the upper $100(1-\alpha)$ percentage point of the χ^2_p distribution. The confidence region is an ellipsoid centred at $\mu = \bar{x}$.

(ii) Σ unknown. When Σ is unknown, the pivotal quantity is $n(\bar{x}-\mu)'S^{-1}(\bar{x}-\mu)$, which we have seen to have a $T^2_p(n-1)$ distribution. Using the equivalent $F_{p,n-p}$ distribution in place of $T^2_p(n-1)$, the $100(1-\alpha)\%$ confidence region for μ is

$$\left\{\mu : n(\mu-\bar{x})'S^{-1}(\mu-\bar{x}) \leq \frac{p(n-1)}{n-p}F_{p,n-p,\alpha}\right\}.$$

Again, the confidence region is an ellipsoid centred at $\mu = \bar{x}$, but will be a larger region than that for the case where Σ is known. Although these confidence regions for μ were based on the multivariate normal distribution, they are appropriate in most cases, because of the central limit theorem.

Example

Using the blood pressure data again, let $x_{pre} = (X_1, X_2)'$ and $x_{post} = (X_3, X_4)'$, with $E(x_{pre}) = \mu_{pre}$, $E(x_{post}) = \mu_{post}$, $var(x_{pre}) = \Sigma_{pre}$ and $var(x_{post}) = \Sigma_{post}$. A 95% confidence region for μ_{pre} is given by

$$15\left[\binom{\mu_1}{\mu_2} - \binom{176.9}{112.3}\right]'\binom{422.9 \quad 143.2}{143.2 \quad 109.7}^{-1}\left[\binom{\mu_1}{\mu_2} - \binom{176.9}{112.3}\right]$$
$$< \frac{2 \times 14}{13}F_{2,13,0.05}.$$

From the F-tables, $F_{2,13,0.05} = 3.81$, and upon expansion, the above inequality becomes

$$0.0042(\mu_1 - 176.9)^2 - 0.0111(\mu_1 - 176.9)(\mu_2 - 112.3) + 0.0163(\mu_2 - 112.3)^2$$
$$< 0.05471.$$

This is an ellipse centred about $(176.9, 112.3)'$, and is shown in Figure 3.3, together with the confidence ellipse for μ_{post}.

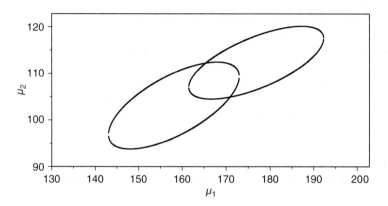

Figure 3.3 Confidence ellipses for μ_{pre} and μ_{post}

3.4.3 Hypothesis testing

Analogous to the use of confidence regions in the estimation of unknown parameters is the testing of hypotheses about the parameters. We will cover only the simple case of testing the mean vector for the multivariate normal distribution. First we test whether the population mean vector, $\boldsymbol{\mu}$, is equal to a particular constant vector, $\boldsymbol{\mu}_0$.

(i) Σ known.   We form the following null and alternative hypotheses

$$H_0: \boldsymbol{\mu} = \boldsymbol{\mu}_0$$
$$H_1: \boldsymbol{\mu} \neq \boldsymbol{\mu}_0.$$

The test statistic is $(\bar{\mathbf{x}} - \boldsymbol{\mu}_0)'\boldsymbol{\Sigma}^{-1}(\bar{\mathbf{x}} - \boldsymbol{\mu}_0)$, and we reject H_0 in favour of H_1, if

$$n(\bar{\mathbf{x}} - \boldsymbol{\mu}_0)'\boldsymbol{\Sigma}^{-1}(\bar{\mathbf{x}} - \boldsymbol{\mu}_0) > \chi^2_{n,\alpha},$$

and accept H_0 otherwise.

(ii) Σ unknown.   This time we use the test statistic $(\bar{\mathbf{x}} - \boldsymbol{\mu}_0)'\mathbf{S}^{-1}(\bar{\mathbf{x}} - \boldsymbol{\mu}_0)$, and we reject H_0 in favour of H_1, if

$$n(\bar{\mathbf{x}} - \boldsymbol{\mu}_0)'\mathbf{S}^{-1}(\bar{\mathbf{x}} - \boldsymbol{\mu}_0) > T^2_{p,\alpha}(n - 1),$$

and accept H_0 otherwise.

Now we look at the more common two-sample test. Suppose $\mathbf{x}_1, \ldots, \mathbf{x}_n$ is a random sample from a $N_p(\boldsymbol{\mu}_{\mathbf{x}}, \boldsymbol{\Sigma})$ distribution, and $\mathbf{y}_1, \ldots, \mathbf{y}_m$ is a random sample from a $N_p(\boldsymbol{\mu}_{\mathbf{y}}, \boldsymbol{\Sigma})$ distribution. Note these distributions have the same covariance matrix. The hypotheses now are

$$H_0: \boldsymbol{\mu}_{\mathbf{x}} = \boldsymbol{\mu}_{\mathbf{y}}$$
$$H_1: \boldsymbol{\mu}_{\mathbf{x}} \neq \boldsymbol{\mu}_{\mathbf{y}}.$$

The test statistic is $nm(n + m - 2)^{-1}(\bar{\mathbf{x}} - \bar{\mathbf{y}})'\mathbf{S}_c^{-1}(\bar{\mathbf{x}} - \bar{\mathbf{y}})$, where

$$\mathbf{S}_c = (n + m - 2)^{-1}[(n - 1)\mathbf{S}_{\mathbf{x}} + (m - 1)\mathbf{S}_{\mathbf{y}}],$$

with $\mathbf{S}_{\mathbf{x}}$ and $\mathbf{S}_{\mathbf{y}}$ being the sample covariance matrices of the respective samples. We reject H_0 in favour of H_1, if

$$\frac{nm}{n + m - 2}(\bar{\mathbf{x}} - \bar{\mathbf{y}})'\mathbf{S}_c^{-1}(\bar{\mathbf{x}} - \bar{\mathbf{y}}) > T^2_{p,\alpha}(n + m - 2),$$

and accept H_0 otherwise.

Example

A person with blood pressure above 160/95 (systolic/diastolic) could be said to have hypertension. Based on this, we test the hypotheses:

$$H_0: \mu_{pre} = \begin{pmatrix} 160 \\ 95 \end{pmatrix}$$

$$H_1: \mu_{pre} \neq \begin{pmatrix} 160 \\ 95 \end{pmatrix}.$$

Now $n(\bar{x} - \mu_0)'S^{-1}(\bar{x} - \mu_0)$ is given by

$$15 \left[\begin{pmatrix} 176.9 \\ 112.3 \end{pmatrix} - \begin{pmatrix} 160 \\ 95 \end{pmatrix} \right]' \begin{pmatrix} 422.9 & 143.2 \\ 143.2 & 109.7 \end{pmatrix}^{-1} \left[\begin{pmatrix} 176.9 \\ 112.3 \end{pmatrix} - \begin{pmatrix} 160 \\ 95 \end{pmatrix} \right]$$

$$= 43.0.$$

Using the F-tables, $T_2^2(14) = \frac{14 \times 2}{14 - 2 + 1} F_{2,14,0.01} = 8.05$, and thus the null hypothesis is emphatically rejected in favour of the alternative. (Looking at the data, this is hardly surprising!)

3.5 Exercises

1. For the blood pressure data, use a computer package that will carry out matrix algebra to find the sample mean vector, the sample covariance matrix, and the sample correlation matrix, using the data matrix X in the equations

$$\bar{x} = n^{-1}X'1$$

$$S = (n-1)^{-1}X'HX$$

$$R = D^{-1}SD^{-1}.$$

2. Let x be the random vector for the blood pressure data as before. Consider the transformation $y = A'x$, where

$$A = \begin{pmatrix} 1 & 1 & 0 & 1 \\ 1 & -1 & 0 & -1 \\ 1 & 0 & 1 & -1 \\ 1 & 0 & -1 & 1 \end{pmatrix}.$$

Interpret the meaning of Y_1, Y_2, Y_3 and Y_4. Find the sample mean vector and sample covariance matrix of y. Hence find the sample correlation matrix of y.

3. Suppose for the blood pressure data that $x \sim N_4(\mu, \Sigma)$. Let $x_1 = (X_1, X_2)'$ be the pre measurements, and $x_2 = (X_3, X_4)'$ be the post measurements. Estimating μ and Σ by the sample mean vector and sample covariance matrix, find $E(x_2|x_1)$ in terms of x_1. Find $var(x_2|x_1)$.

4. Find a 95% confidence region for μ_{post} for the blood pressure data.

5. For the blood pressure data test the following hypotheses

$$H_0: \mu_{pre} = \mu_{post}$$
$$H_1: \mu_{pre} \neq \mu_{post}.$$

6. Download the dataset 'Brain size' from the DASL website. The data consist of three IQ scores (FSIQ, VIQ, PIQ), weight, height, and an MRI count that measures brain size, for a sample of right-handed psychology students. The data were supplied by Willerman et al. (1991).

Let $\mathbf{x} = (FSIQ, VIQ, PIQ)'$ for males, and $\mathbf{y} = (FSIQ, VIQ, PIQ)'$ for females. Calculate the sample mean vectors and sample covariance matrices for $\mathbf{x}$ and $\mathbf{y}$. Test for a differences between mean scores for males and females.

4

Graphical representation of multivariate data

A golden rule of statistics is to summarize data, whenever possible, with a graph, before attempting a complex statistical analysis. In addition, results of subsequent analyses are often best illustrated by graphical means. For univariate data, graphical representation is easily accomplished using bar charts, histograms, pie charts, stem and leaf plots, and other such methods. Nowadays, statistical software packages can produce wonderful, colourful graphical illustrations of univariate data. However, plotting multivariate data is not so easy, unless there are only two variables. Some ingenious methods for more than two dimensions have been devised and many statistical software packages can produce these plots. Most are designed to be interactive so that the analyst can quickly explore various aspects of the data at the touch of a mouse button. In this chapter we illustrate some of the techniques, using S-Plus to produce most of the plots. Here, we can only illustrate static non-interactive plots in black and white, but the real power of modern multivariate graphical techniques is their interactive coloured displays, as well as animation with spinning and dynamic plots.

Four datasets will be used in this chapter to illustrate various graphical representations of data. The first consists of seven variables from the Coronary Risk-Factor Study carried out in South Africa (Rousseauw et al., 1983). The second consists of nine variables relating to patients with prostate cancer. The data come from a study by Stamey et al. (1989). Both datasets are used in Hastie et al. (2001), and can be found at the website http://www-stat.stanford.edu/~tibs/ElemStatLearn/. The third dataset comes from the archaeo-logical database held at the University of Venice (http://venus.unive.it/termo/DataBank/BancaDati.htm) giving the chemical composition of pottery artefacts. The data used here are artefacts from Chania in Crete. The fourth dataset consists of eighteen measurements for bodyfat on 252 men. The data are courtesy of Dr. A Garth Fisher, and can be found at the website: http://www.amstat.org/publications/jse/jse_data_archive.html.

The starting point for bivariate data is a scatterplot. Figure 4.1 shows a scatterplot of adiposity against age for the heart disease data. Immediately, it can be seen that adiposity increases with age. However, a word of caution is necessary here, because viewing just two dimensions of a ten-dimensional dataset can miss the true underlying structure within the data. To illustrate this, Figure 4.2 shows separate one-dimensional plots of adiposity and age. From these it is impossible to infer the relationship between adiposity and age as seen in Figure 4.1 (i.e. low adiposity values are associated with low age values; high adiposity

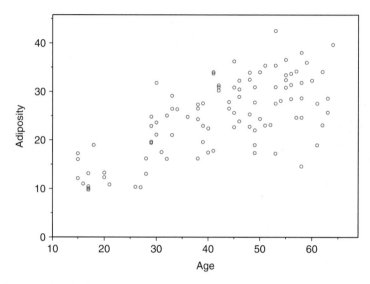

Figure 4.1 Scatterplot of adiposity against age

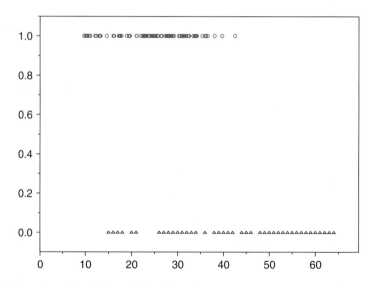

Figure 4.2 One-dimensional plots of adiposity and age

values are associated with high age values). The same happens for our two-dimensional scatterplots: they cannot tell us about the overall pattern in ten dimensions. Nevertheless, we must still use them as they are the best we have.

A scatterplot matrix is a matrix of scatterplots between all pairs of variables. Figure 4.3 shows a scatterplot matrix of the Chania data. It is left to the reader to peruse the 2-D relationships between the variables.

It is possible to add extra dimensions to 2-D scatterplots, for instance by using colour or by varying the size of the symbols depicting the points. Figure 4.4 shows a scatterplot of weight against neck circumference for the bodyfat, but where the size of the circle depends

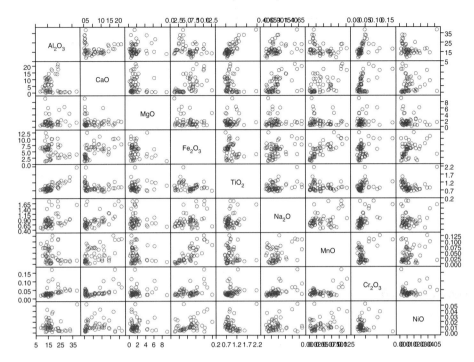

Figure 4.3 Scatterplot matrix of the Chania data

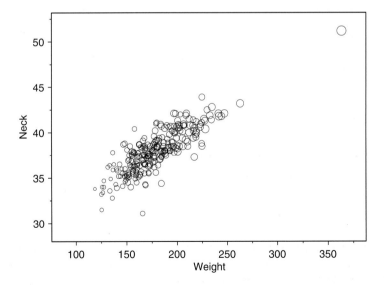

Figure 4.4 Bubble plot of weight, neck and thigh

on thigh circumference. This is called a bubble plot. It can be seen that neck circumference increases with weight, but so does thigh circumference.

Figure 4.5 shows a contour plot of Aindex over the values of age and wrist circumference for the bodyfat data. The higher areas of Aindex appear to the top and to the right of the plot.

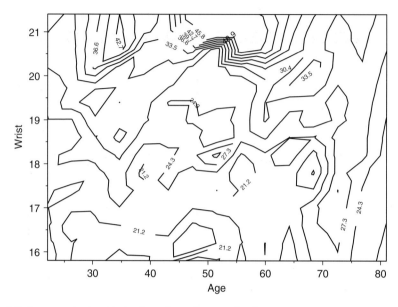

Figure 4.5 Contour plot for Aindex for values of age and wrist

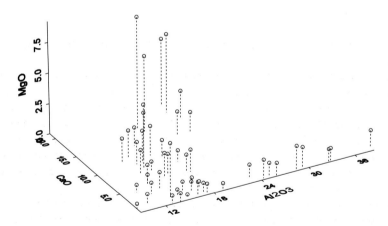

Figure 4.6 Three-dimensional scatterplot of the Chania data

It is possible to produce a three-dimensional scatterplot for three variables, usually with a device for giving perspective to the plot. Figure 4.6 shows a three-dimensional scatterplot for MgO, CaO and Al_2O_3 for the Chania data. There is an interesting grouping of the points, where for one group, MgO, CaO and Al_2O_3 increase together, while for another, MgO and CaO do not increase as Al_2O_3 increases.

Sometimes it is useful to plot a series of scatterplots for two variables, where each is conditioned on another variable or perhaps several variables. Trellis graphics in S-Plus does this. Figure 4.7 shows CaO plotted against Al_2O_3 for the four intervals (0.014, 0.024), (0.024, 0.030), (0.031, 0.041) and (0.041, 0.226) covering the range of Cr_2O_3. An interesting split

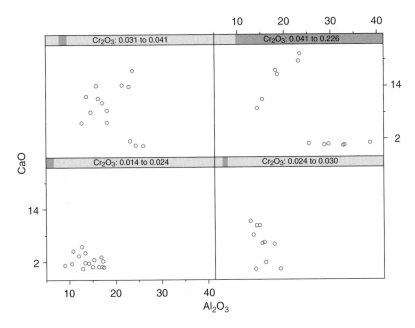

Figure 4.7 Trellis graphics for CaO, Al$_2$O$_3$ conditioned on Cr$_2$O$_3$

in the data can be seen for the final interval, and also for the penultimate interval, but not so markedly.

Figure 4.8 shows a parallel axes plot for the prostate data. Each variable has an axis and each observation is marked as a point on each of the axes. These points are then joined from top to bottom for each of the observations. For many data points the plot can become confused, but hopefully structure within the data is highlighted. The data used in Figure 4.8 are the prostate data, split into two groups, one for the patients where there was seminal vesicle invasion (svi $= 1$) and one where there was not (svi $= 0$). The plot is more effective in colour. Differences between the two plots can be seen, with the variables lcavol, lcp and lpsa standing out in particular.

Figure 4.9 shows a star plot for the first sixteen patients from the prostate cancer data. One star is plotted for each observation. Each variable has an axis that radiates from the origin. The axes are drawn sequentially at a set number of degrees, until the circle is completed. The data are scaled appropriately and then, for each variable, a line is drawn along the appropriate axis, of length proportional to the actual observation. The endpoints of the lines are then joined.

In an Andrews' plot, each data vector, x_r, is plotted as a harmonic curve

$$f_r(t) = x_{r1}/\sqrt{2} + x_{r2} \sin t + x_{r3} \cos t + x_{r4} \sin 2t + x_{r5} \cos 2t$$
$$+ x_{r6} \sin 3t + x_{r7} \cos 3t + \cdots$$

the sum ending when all the variables have been used up. Then $f_r(t)$ is plotted against t, for $\pi \leq t \leq \pi$. The curve is composed of a sum of harmonics, with amplitudes related to the various variables. The first variable, x_1, gives the level about which f_r oscillates.

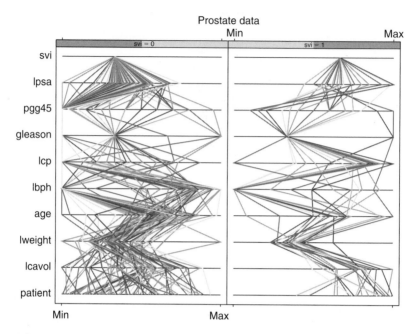

Figure 4.8 Parallel axes plots for the prostate cancer data

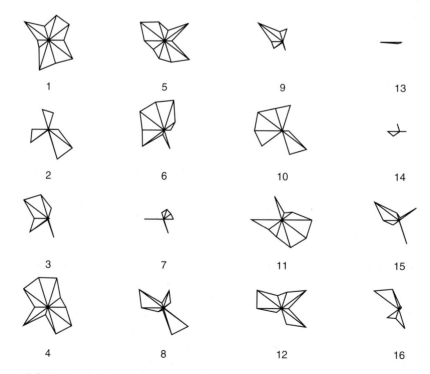

Figure 4.9 Star plot for the prostate cancer data

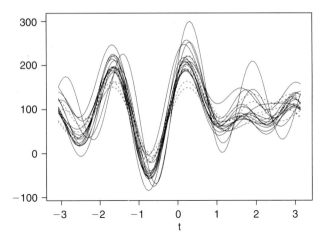

Figure 4.10 Andrews' plot for the heart disease data

The amplitude, A_1, and phase, ϕ_1, of the lowest frequency harmonic, $A_1 \sin(t + \phi_1)$, are given by $\sqrt{x_{r2}^2 + x_{r3}^2}$, and $\sin^{-1}\left(x_{r3}/\sqrt{x_{r2}^2 + x_{r3}^2}\right)$, with similar results for the other harmonics $A_2 \sin(2t + \phi_2)$, etc. The functions $f_r(t)$, $r = 1, \ldots, n$, are plotted together, in the hope that visual inspection will indicate groupings of the observations and perhaps outliers.

Figure 4.10 shows an Andrews' plot for twenty of the observations for the heart disease data. The solid lines are plots for males who were over forty years of age, and the dotted lines are plots for males under forty years of age. This latter group have less variability in their plots.

One reason for this particular choice of $f_r(t)$ is that it preserves the Euclidean distances between data vectors (observations), where distance between the two curves f_r and f_s is defined as

$$\int (f_r(t) - f_s(t))^2 dt.$$

One unfortunate aspect of Andrews' curves is that their shape is greatly affected by the order in which the variables enter the function f.

One interesting and often amusing technique for illustrating multivariate data is the use of Chernoff's faces. Each variable is related to a particular feature of the face. For instance, x_1 may be related to the overall size of the face, the larger the value, the larger the face. The variable x_2 may be related to the size of the nose, x_3 to the displacement of the eyes, etc. Figure 4.11 shows Chernoff faces for the prostate cancer data, where *age* is represented by area of the face, *gleason* by the shape of head, *lcavol* by the length of the nose, *lweight* by the location of the mouth, *lbph* by the curve of the smile, *lcp* by the width of the mouth, *pgg45* by the location of the eyes, and *lpsa* by the separation of the eyes. Once plotted, 'families' of faces are sought, where faces appear similar. For instance, faces 1, 2, 3, 4 and 10 might be chosen as a family who share similar characteristics.

Overall, the graphical illustration of multivariate data is very useful for an initial investigation; see Basford and Tukey (1998) for further coverage. However, graphical illustration

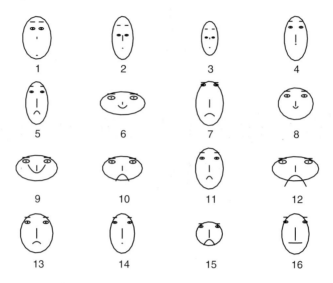

Figure 4.11 Chernoff's faces for the prostate data

is no substitute for actual multivariate analysis by the various methods discussed in the subsequent chapters.

4.1 Exercises

1. Choose a dataset with a relatively low number of observations and variables ($n = 20$, $p = 5$ say), and use whichever statistical software package(s) you have at hand to illustrate the data graphically.

   Now choose a large dataset with many observations and variables, and illustrate the data graphically. Which methods are impracticable for large datasets?

2. The squared Euclidean distance between two observations is given by

$$\sum_{i=1}^{p} (x_{ri} - x_{si})^2.$$

Show that Andrews' curves preserve this Euclidean distance. You may find the following helpful:

$$\int_{-\pi}^{\pi} \sin it \sin jt \, dt = 0 \quad (i \neq j)$$

$$\int_{-\pi}^{\pi} \cos it \cos jt \, dt = 0 \quad (i \neq j)$$

$$\int_{-\pi}^{\pi} \sin it \cos jt \, dt = 0.$$

5

Principal components analysis

Principal components analysis (PCA) is a dimension reduction technique for quantitative data. Suppose twenty measurements are made on a set of prehistoric artefacts, giving us data in twenty dimensions. Would it be possible to reduce this number of dimensions down to two or three, but at the same time, keeping as much information as to the inherent variability within the data? This is the aim of principal components analysis. In general, let the variables measured on each artefact, object, or individual be denoted by $X_1, \ldots X_p$, (the 'original variables'). The aim is to find new variables, $Y_1, \ldots, Y_p$, which are: (i) linear combinations of the original variables, (ii) uncorrelated with each other, and are such that, (iii) just a few of them will explain most of the variation in the data, hence effectively reducing the number of dimensions. Principal components analysis is a completely general method which does not assume any model for the data, such as multivariate normality. Of course, although the method can be used on any data, even categorical data if coded appropriately, there is no guarantee that results will be meaningful.

Books that cover the theory of principal components analysis in detail are Jackson (1991) and Jolliffe (2002).

5.1 Theory

Let $\mathbf{x} = (X_1, X_2, \ldots, X_p)'$ denote the original variables measured on the objects or individuals. Let $\mathbf{y} = (Y_1, Y_2, \ldots, Y_p)'$ denote the derived set of variables. Principal components analysis transforms $\mathbf{x}$ to $\mathbf{y}$, such that

(i) $Y_j = a_{1j}X_1 + a_{2j}X_2 + \ldots + a_{pj}X_p \quad (j = 1, \ldots, p)$

(ii) $\mathrm{corr}(Y_j, Y_k) = 0 \quad (j \neq k)$

(iii) the $Y_j's$ are labelled so that $\mathrm{var}(Y_1) \geq \mathrm{var}(Y_2) \geq \ldots \geq \mathrm{var}(Y_p)$.

The derived variable Y_1 is called the first principal component (PC), Y_2 is the second principal component, and so on.

Let $\mathbf{x}$ have mean vector $\boldsymbol{\mu}$ and covariance matrix $\boldsymbol{\Sigma}$. Let $\mathbf{a}_j = (a_{1j}, \ldots, a_{pj})'$. The principal components are

$$Y_j = \mathbf{a}_j'\mathbf{x}.$$

The first principal component, Y_1, is found by choosing $\mathbf{a}_1$, such that Y_1 has the largest possible variance. We could choose $\mathbf{a}_1$ so that var(Y_1) is infinite, and so to prevent this degenerate case, we impose the scaling, $\mathbf{a}_1'\mathbf{a}_1 = 1$, and similarly for all the principal components, $\mathbf{a}_j'\mathbf{a}_j = 1$ $(j = 1, \ldots, p)$.

Now var(Y_1) = var($\mathbf{a}_1'\mathbf{x}$) = $\mathbf{a}_1'\boldsymbol{\Sigma}\mathbf{a}_1 = V$ (say). We maximize var(Y_1) using a Lagrange multiplier, λ, and so consider

$$V = \mathbf{a}_1'\boldsymbol{\Sigma}\mathbf{a}_1 - \lambda(\mathbf{a}_1'\mathbf{a}_1 - 1).$$

Differentiating with respect to $\mathbf{a}_1$, and equating to $\mathbf{0}$ gives

$$\frac{\partial V}{\partial \mathbf{a}_1} = 2\boldsymbol{\Sigma}\mathbf{a}_1 - 2\lambda\mathbf{a}_1 = \mathbf{0}.$$

Hence

$$(\boldsymbol{\Sigma} - \lambda\mathbf{I})\mathbf{a}_1 = \mathbf{0}. \tag{5.1}$$

In order to have a solution, apart from $\mathbf{a}_1 = \mathbf{0}$, the matrix $(\boldsymbol{\Sigma} - \lambda\mathbf{I})$ must be a singular matrix and hence its determinant is zero. Thus λ satisfies the equation

$$|\boldsymbol{\Sigma} - \lambda\mathbf{I}| = 0,$$

and hence we have a solution, if and only if λ is an eigenvalue of $\boldsymbol{\Sigma}$.

Now $\boldsymbol{\Sigma}$ has p eigenvalues, which are all non-negative, since $\boldsymbol{\Sigma}$ is positive semi-definite. Let these eigenvalues be $\lambda_1, \lambda_2, \ldots, \lambda_p$, and for the moment, assume these eigenvalues are distinct, and labelled such that $\lambda_1 > \lambda_2 > \ldots > \lambda_p \geq 0$. Now

$$\begin{aligned}
\text{var}(Y_1) = \text{var}(\mathbf{a}_1'\mathbf{x}) &= \mathbf{a}_1'\boldsymbol{\Sigma}\mathbf{a}_1 \\
&= \mathbf{a}_1'\lambda\mathbf{I}\mathbf{a}_1 \quad \text{(from (5.1))} \\
&= \lambda\mathbf{a}_1'\mathbf{a}_1 \\
&= \lambda.
\end{aligned}$$

Since var(Y_1) is to have maximum variance, λ has to be chosen as the largest eigenvalue, λ_1, and hence $\mathbf{a}_1$ must be the right eigenvector of $\boldsymbol{\Sigma}$ corresponding to λ_1.

The second principal component, Y_2, is found in a similar way. We have $Y_2 = \mathbf{a}_2'\mathbf{x}$, with the normalizing condition, $\mathbf{a}_2'\mathbf{a}_2 = 1$. We also need cov($Y_1, Y_2$) = 0. Now

$$\begin{aligned}
\text{cov}(Y_1, Y_2) &= \text{cov}(\mathbf{a}_2'\mathbf{x}, \mathbf{a}_1'\mathbf{x}) \\
&= E[\mathbf{a}_2'(\mathbf{x} - \boldsymbol{\mu}) \times \mathbf{a}_1'(\mathbf{x} - \boldsymbol{\mu})] \\
&= E[\mathbf{a}_2'(\mathbf{x} - \boldsymbol{\mu})(\mathbf{x} - \boldsymbol{\mu})'\mathbf{a}_1]
\end{aligned}$$

$$= \mathbf{a}_2' E[(\mathbf{x} - \boldsymbol{\mu})(\mathbf{x} - \boldsymbol{\mu})']\mathbf{a}_1$$
$$= \mathbf{a}_2' \boldsymbol{\Sigma} \mathbf{a}_1$$
$$= \mathbf{a}_2' \lambda_1 \mathbf{a}_1 = \lambda_1 \mathbf{a}_2' \mathbf{a}_1$$
$$= 0, \quad \text{if and only if,} \quad \mathbf{a}_2' \mathbf{a}_1 = 0.$$

Thus $\mathbf{a}_1$ and $\mathbf{a}_2$ must be orthogonal.

We now maximize var(Y_2) with two constraints,

$$V = \text{var}(Y_2) = \mathbf{a}_2' \boldsymbol{\Sigma} \mathbf{a}_2 - \lambda(\mathbf{a}_2' \mathbf{a}_2 - 1) - \delta(\mathbf{a}_2' \mathbf{a}_1 - 0).$$

Differentiating with respect to $\mathbf{a}_2$, and equating to $\mathbf{0}$,

$$\frac{\partial V}{\partial \mathbf{a}_2} = 2(\boldsymbol{\Sigma} - \lambda \mathbf{I})\mathbf{a}_2 - \delta \mathbf{a}_1 = \mathbf{0}. \tag{5.2}$$

Premultiply by $\mathbf{a}_1'$, to give

$$2\mathbf{a}_1'(\boldsymbol{\Sigma} - \lambda \mathbf{I})\mathbf{a}_2 - \delta \mathbf{a}_1' \mathbf{a}_1 = 0,$$

i.e.

$$2\mathbf{a}_1' \boldsymbol{\Sigma} \mathbf{a}_2 - 2\lambda \mathbf{a}_1' \mathbf{a}_2 - \delta \mathbf{a}_1' \mathbf{a}_1 = 0. \tag{5.3}$$

But $\mathbf{a}_1' \boldsymbol{\Sigma} \mathbf{a}_2 = \text{cov}(Y_1, Y_2) = 0$, $\mathbf{a}_1' \mathbf{a}_2 = 0$ and $\mathbf{a}_1' \mathbf{a}_1 = 1$, and thus equation (5.3) reduces to

$$\delta = 0.$$

Hence equation (5.2) becomes

$$(\boldsymbol{\Sigma} - \lambda \mathbf{I})\mathbf{a}_2 = \mathbf{0},$$

and

$$\text{var}(Y_2) = \text{var}(\mathbf{a}_2' \mathbf{x}) = \mathbf{a}_2' \boldsymbol{\Sigma} \mathbf{a}_2 = \mathbf{a}_2' \lambda \mathbf{a}_2 = \lambda.$$

Thus, this second λ is also an eigenvalue of $\boldsymbol{\Sigma}$. It cannot be chosen as the largest eigenvalue, since that would lead us to the first principal component again. It has to be chosen as the second largest eigenvalue, and then $\mathbf{a}_2$ is the right eigenvector associated with it. This has given us the first two principal components. The process is continued in order to obtain the rest of the principal components, $Y_3, Y_4, \ldots, Y_p$. It is found that the jth principal component is given by $Y_j = \mathbf{a}_j' \mathbf{x}$, where $\mathbf{a}_j$ is the jth eigenvector of $\boldsymbol{\Sigma}$, and var$(Y_j) = \lambda_j$, where λ_j is the jth eigenvalue of $\boldsymbol{\Sigma}$.

If some of the eigenvalues are equal, the above argument can be extended. However, there is no unique way of choosing the eigenvectors of multiple roots. We ensure these eigenvectors are chosen to be orthogonal and the arguments carry through.

We now put all the principal components into one vector, $\mathbf{y} = (Y_1, Y_2, \ldots, Y_p)'$. Put the eigenvectors of $\boldsymbol{\Sigma}$ into matrix $\mathbf{A}$, so that $\mathbf{A} = [\mathbf{a}_1, \mathbf{a}_2, \ldots, \mathbf{a}_p]$. Then

$$\mathbf{y} = \mathbf{A}' \mathbf{x},$$

with

$$\text{var}(\mathbf{y}) = \boldsymbol{\Lambda} = \text{diag}(\lambda_1, \lambda_2, \dots, \lambda_p).$$

Now

$$\text{var}(\mathbf{y}) = \boldsymbol{\Lambda} = \mathbf{A}'\boldsymbol{\Sigma}\mathbf{A},$$

and so, as $\mathbf{A}$ is orthogonal with $\mathbf{A}^{-1} = \mathbf{A}'$,

$$\boldsymbol{\Sigma} = \mathbf{A}\boldsymbol{\Lambda}\mathbf{A}',$$

showing that finding principal components is essentially the same as finding the spectral decomposition of $\boldsymbol{\Sigma}$.

Now

$$\sum_{j=1}^{p} \text{var}(Y_j) = \sum_{j=1}^{p} \lambda_j = \text{tr}(\boldsymbol{\Lambda})$$

$$= \text{tr}(\mathbf{A}'\boldsymbol{\Sigma}\mathbf{A})$$

$$= \text{tr}(\boldsymbol{\Sigma}\mathbf{A}\mathbf{A}')$$

$$= \text{tr}(\boldsymbol{\Sigma}).$$

Therefore

$$\sum_{j=1}^{p} \text{var}(Y_j) = \sum_{j=1}^{p} \text{var}(X_j),$$

and so the sum of the variances of the principal components equals the sum of the variances of the original variables. We say that $\sum_{j=1}^{p} \text{var}(X_j)$ is the total variation in the data, and that the jth principal component accounts for the proportion

$$\frac{\lambda_j}{\sum_{i=1}^{p} \lambda_i}$$

of this. Similarly, the first m principal components account for

$$\frac{\sum_{i=1}^{m} \lambda_i}{\sum_{i=1}^{p} \lambda_i}$$

of the total variation.

Example

Let X_1, X_2, X_3 have the covariance matrix

$$\boldsymbol{\Sigma} = \begin{bmatrix} 13 & 1 & 4 \\ 1 & 13 & 4 \\ 4 & 4 & 10 \end{bmatrix}.$$

To find the principal components, we first find the eigenvalues and eigenvectors of Σ. Solving

$$|\Sigma - \lambda\mathbf{I}| = \begin{vmatrix} 13-\lambda & 1 & 4 \\ 1 & 13-\lambda & 4 \\ 4 & 4 & 10-\lambda \end{vmatrix} = 0,$$

gives eigenvalues $\lambda = 18, 12, 6$, and associated eigenvectors $(1/\sqrt{3}, 1/\sqrt{3}, 1/\sqrt{3})'$, $(1/\sqrt{2}, -1/\sqrt{2}, 0)'$ and $(1/\sqrt{6}, 1/\sqrt{6}, -2/\sqrt{6})'$, noting that the eigenvectors have been standardized to have length unity. So, the first principal component is given by

$$Y_1 = \frac{1}{\sqrt{3}}X_1 + \frac{1}{\sqrt{3}}X_2 + \frac{1}{\sqrt{3}}X_3,$$

with $\text{var}(Y_1) = 18$. The second principal component is

$$Y_2 = \frac{1}{\sqrt{2}}X_1 - \frac{1}{\sqrt{2}}X_2,$$

with $\text{var}(Y_2) = 12$, and the third principal component is

$$Y_3 = \frac{1}{\sqrt{6}}X_1 + \frac{1}{\sqrt{6}}X_2 - \frac{2}{\sqrt{6}}X_3,$$

with $\text{var}(Y_3) = 6$. The first principal component explains $100 \times 18/(18+12+6) = 50\%$ of the total variation, the second explains $100 \times 12/(18+12+6) = 33.3\%$ of the total variation, and the third, $100 \times 6/(18+12+6) = 16.7\%$ of the total variation. The first PC is essentially a mean of the three X variables, the second 'contrasts' X_1 against X_2, and the third contrasts X_1 and X_2 against X_3.

Note that if $\mathbf{a}_i$ is the eigenvector associated with λ_i, the ith eigenvalue of Σ, and giving the ith PC as $\mathbf{a}_i'\mathbf{x}$, then $-\mathbf{a}_i'\mathbf{x}$ would also serve as the ith PC. This is because the signs of eigenvectors are arbitrary: $\Sigma\mathbf{a}_i = \lambda_i\mathbf{a}_i$ is equivalent to $\Sigma(-\mathbf{a}_i) = \lambda_i(-\mathbf{a}_i)$. The choice as to which sign is used has to be made by the analyst.

So far, we have assumed that Σ is known. This is usually not the case, and Σ has to be estimated by the sample covariance matrix, $\mathbf{S}$, giving rise to 'estimated principal components'. However, we usually do not speak of estimated principal components, but simply refer to them as principal components (or PC's).

Example: PCA of bodyfat data

We use the bodyfat data introduced in Chapter 4. Percentage of bodyfat, age, weight, height and ten body circumference measurements were measured on 252 men. We use the circumference measurements (neck, chest, abdomen, hip, thigh, knee, ankle, biceps, forearm and wrist), all measured in centimetres, to illustrate the technique of principal components analysis. Table 5.1 shows the sample mean and standard deviations for each of the variables.

Table 5.1 Mean and sd of body circumference measurements

Measurement	$\bar{x}$	s
neck	38.0	2.43
chest	100.8	8.43
abdomen	92.6	10.78
hip	99.9	7.16
thigh	59.4	5.25
knee	38.6	2.41
ankle	23.1	1.70
biceps	32.3	3.02
forearm	28.7	2.02
wrist	18.2	0.93

The sample covariance matrix of the ten variables, where only the upper triangle of the matrix is shown, is

	neck	chest	abdo.	hip	thigh	knee	ankle	biceps	f'arm	wrist
neck	5.91	16.08	19.77	12.80	8.88	3.94	1.97	5.37	3.06	1.69
chest	.	71.07	83.25	50.09	32.30	14.63	6.90	18.54	9.88	5.20
abdomen	.	.	116.3	67.52	43.40	19.17	8.28	22.32	10.97	6.24
hip	.	.	.	51.32	33.71	14.23	6.78	16.00	7.89	4.21
thigh	.	.	.	.	27.56	10.11	4.80	12.08	6.01	2.74
knee	.	.	.	.	.	5.82	2.50	4.95	2.71	1.50
ankle	.	.	.	.	.	.	2.87	2.48	1.44	0.90
biceps	.	.	.	.	.	.	.	9.13	4.14	1.78
forearm	.	.	.	.	.	.	.	.	4.08	1.10
wrist	.	.	.	.	.	.	.	.	.	0.87

The sum of the ten variances in the covariance matrix is 294.9.

The eigenvalues and eigenvectors of the covariance matrix, giving rise to the principal components, are

E'value	255.7	16.53	8.80	4.00	2.87	2.25	1.70	1.57	1.23	0.28
PC:	PC1	PC2	PC3	PC4	PC5	PC6	PC7	PC8	PC9	PC10
neck	0.12	−0.02	0.20	0.24	−0.26	−0.07	0.62	0.62	0.04	0.20
chest	0.50	0.38	0.64	−0.36	0.24	−0.01	−0.03	−0.02	0.02	0.00
abdomen	0.66	0.38	−0.55	0.33	−0.04	−0.06	−0.04	−0.05	−0.04	0.00
hip	0.42	−0.51	−0.18	−0.52	−0.38	0.34	0.06	−0.03	−0.04	0.01
thigh	0.28	−0.60	0.02	0.22	0.68	−0.18	0.11	0.02	−0.09	−0.06
knee	0.12	−0.18	0.04	−0.01	−0.20	−0.52	−0.17	−0.09	0.77	0.11
ankle	0.06	−0.12	0.10	−0.06	−0.29	−0.60	−0.35	0.14	−0.61	0.12
biceps	0.15	−0.18	0.34	0.51	−0.18	0.44	−0.55	0.17	0.07	0.03
forearm	0.07	−0.09	0.29	0.33	−0.29	−0.03	0.37	−0.74	−0.15	0.05
wrist	0.04	−0.01	0.08	0.05	−0.18	−0.12	0.06	0.10	0.02	−0.96

Hence, the first principal component is

$$Y_1 = 0.12X_1 + 0.50X_2 + 0.66X_3 + 0.42X_4 + 0.28X_5 + 0.12X_6$$
$$+ 0.06X_7 + 0.15X_8 + 0.07X_9 + 0.04X_{10},$$

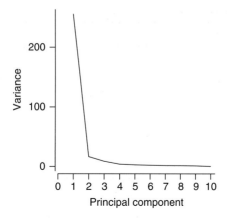

Figure 5.1 Scree plot for bodyfat PCs

and since all the coefficients are positive, it can be seen as an overall measurement of 'size'. The variables contributing most to Y_1 are chest, abdomen, hip and thigh. The second principal component is

$$Y_2 = -0.02X_1 + 0.38X_2 + 0.38X_3 - 0.51X_4 - 0.60X_5 - 0.18X_6$$
$$- 0.12X_7 - 0.18X_8 - 0.09X_9 - 0.01X_{10}.$$

Looking at the largest positive and largest negative values of the coefficients, we see that the second principal component is essentially contrasting chest and abdomen circumferences with those of the hip and thigh. The other principal components can be looked at in the same manner, but noting their importance according to the amount of variation they explain.

Figure 5.1 shows a plot of the eigenvalues/variances for the principal components from the first to the last. This is called a *scree plot*, and helps in deciding the number of principal components to choose in order to represent the original variables. From the plot, and the table containing the eigenvalues, it is clear that the first principal component accounts for the majority of the variation in the data. It accounts for $100 \times 255.7/(255.7 + \ldots + 0.28) = 86.7\%$ of the total variation. Between them, the first two principal components account for 92.3% of the total variation.

5.2 Problems

One problem with PCA is that it is not scale invariant. If the scale of measurement is changed for one of the variables, then the results from the analysis change. If the original variables are all measured on the same scale and are of similar magnitude, then the analysis will probably be sound. Otherwise, it is usually best to find principal components from the correlation matrix, rather than the covariance matrix. This puts the original variables on the same footing.

Another problem is one of zero eigenvalues. A zero eigenvalue of $\mathbf{\Sigma}$ or $\mathbf{P}$ implies linear dependence of some of the variables. When using the sample covariance matrix, $\mathbf{S}$, or the sample correlation matrix, $\mathbf{R}$, these zero eigenvalues will not be detected as zero, but as very small eigenvalues. If these occur, then it is worth inspecting the data further to search for these linear dependencies. The larger elements in the eigenvector associated with a small eigenvalue will indicate the variables involved in the linearity.

The problem of repeated eigenvalues of Σ was discussed in the derivation of the principal components. These repeated eigenvalues do not lead to unique eigenvectors and hence to unique principal components. In practice, these repeated eigenvalues will not be detected as being equal when using S, and only the fact that they are close in value will be noted.

Example: Bodyfat continued

Although the ten circumference measurements in the bodyfat dataset are all measured in centimetres, the overall scales of the various variables are disparate. For instance, the maximum abdomen circumference is 148.1 cm, whilst the maximum wrist circumference is 21.4 cm. It may be more insightful to carry out a principal components analysis on the sample correlation matrix, rather than the sample covariance matrix.

The sample correlation matrix for the body circumference data is

	neck	chest	abdo.	hip	thigh	knee	ankle	biceps	f'arm	wrist
neck	1.00	0.79	0.75	0.74	0.70	0.67	0.48	0.73	0.62	0.75
chest	.	1.00	0.92	0.83	0.73	0.72	0.48	0.73	0.58	0.66
abdomen	.	.	1.00	0.87	0.77	0.74	0.45	0.69	0.50	0.62
hip	.	.	.	1.00	0.90	0.82	0.56	0.74	0.55	0.63
thigh	.	.	.	.	1.00	0.80	0.54	0.76	0.57	0.56
knee	.	.	.	.	.	1.00	0.61	0.68	0.56	0.67
ankle	.	.	.	.	.	.	1.00	0.49	0.42	0.57
biceps	.	.	.	.	.	.	.	1.00	0.68	0.63
forearm	.	.	.	.	.	.	.	.	1.00	0.59
wrist	.	.	.	.	.	.	.	.	.	1.00

The principal components are

E'value	7.02	0.73	0.67	0.49	0.30	0.28	0.20	0.16	0.08	0.06
PC:	PC1	PC2	PC3	PC4	PC5	PC6	PC7	PC8	PC9	PC10
neck	0.33	0.00	0.26	0.34	0.05	0.29	0.72	-0.32	-0.08	-0.02
chest	0.34	0.27	0.06	0.24	-0.45	-0.08	-0.24	-0.13	0.54	-0.42
abdomen	0.33	0.40	-0.07	0.22	-0.31	-0.15	-0.13	0.06	-0.30	0.67
hip	0.35	0.26	-0.21	-0.12	0.06	-0.07	0.07	0.35	-0.55	-0.56
thigh	0.33	0.19	-0.18	-0.41	0.26	0.11	0.29	0.40	0.52	0.23
knee	0.33	-0.02	-0.27	-0.14	0.45	-0.44	-0.12	-0.62	0.01	0.01
ankle	0.25	-0.63	-0.58	-0.02	-0.42	0.17	0.07	-0.02	-0.02	0.05
biceps	0.32	-0.02	0.26	-0.30	0.09	0.67	-0.47	-0.20	-0.13	0.03
forearm	0.27	-0.36	0.59	-0.40	-0.26	-0.44	0.09	0.09	-0.07	0.03
wrist	0.30	-0.38	0.14	0.57	0.43	-0.07	-0.27	0.40	0.08	0.03

The first principal component accounts for 70.2% of the total variation, the second, 7.0%, and the rest, 6.7%, 4.9%, 3.0%, 2.8%, 2.0%, 1.6%, 0.8% and 0.6%, respectively. The coefficients in the first principal component are all close to the value 0.32, again making this principal component a measure of size. The second principal component contrasts abdomen circumference with ankle, forearm and wrist circumferences, i.e. body versus limbs. The third principal component contrasts ankle circumference with forearm circumference, i.e. leg versus arm. The fourth principal component contrasts wrist circumference against forearm, biceps and thigh circumference, i.e. arm extremity with limbs, and so forth.

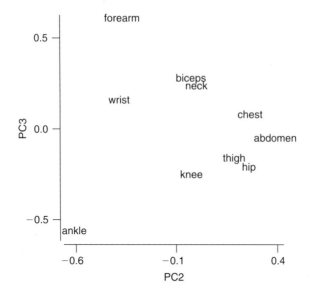

Figure 5.2 Coefficients for PC3 against those for PC2

Figure 5.2 shows the coefficients of PC3 plotted against those of PC2. This type of plot helps to understand which variables have a similar involvement within the PCs. The coefficients for PC1 were not used for plotting here, since they were very similar for all ten variables. A group of variables from the region of the arm (forearm, wrist, biceps, neck) can be seen towards the top left of the plot, a group (chest, abdomen, hip, thigh, knee) towards the lower right, and ankle stands by itself.

5.3 Component scores

The values of the principal components can be calculated for each object or individual. These are called the *component scores*. Let these be placed in a matrix $\mathbf{Y}$, so that the rth row of $\mathbf{Y}$ contains the p component scores for the rth object or individual. Recalling that $\mathbf{A}$ is the matrix of eigenvectors, all the component scores are easily calculated by $\mathbf{Y} = \mathbf{XA}$, or if mean corrected scores are required, by

$$\mathbf{Y} = (\mathbf{X} - \mathbf{1\bar{x}})\mathbf{A}.$$

For the rth individual

$$\mathbf{y}_r = \mathbf{A}'(\mathbf{x}_r - \mathbf{\bar{x}}).$$

Note that if the correlation matrix is being used for PCA, then the data matrix $\mathbf{X}$ has also to be standardized, so that the sample standard deviation of X_i $(i = 1, \ldots, p)$ is unity.

Component scores for the various PC's are then plotted against each other for the individuals, usually only using the first few PCs. Figure 5.3 shows the component scores for the second principal component plotted against those for the first for the bodyfat data, where the principal components have been found from the sample correlation matrix. The individual

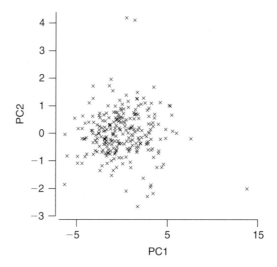

Figure 5.3 Component scores for PC2 plotted against those for PC1

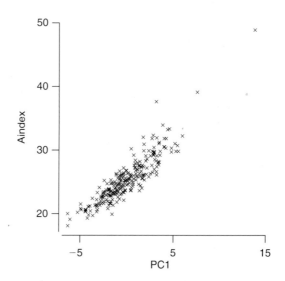

Figure 5.4 Plot of adiposity index against PC1

represented by the point situated to the far right of the plot was the heaviest individual with a weight of 363.15 lbs. The two individuals represented by the points at the very top of the plot are cases 31 and 86. The distinguishing feature for these individuals is that they had the two largest ankle circumferences, 33.7 cm and 33.9 cm.

Adiposity index is a measure of 'well-being', and is defined as weight/height^2, (kg/m^2). Out of interest, this is plotted against the component score for PC1, for each individual in Figure 5.4. There is obviously a very high correlation of the adiposity index with PC1. Figure 5.5 shows the component scores for PC3 plotted against those for PC2.

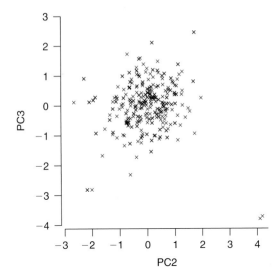

Figure 5.5 Component scores for PC3 plotted against those for PC2

5.4 Component correlations

The principal components are correlated with the original variables. Inspection of the correlations between the principal components and the original variables will show which variables are the most influential in each of the principal components. First, consider the correlation of the jth principal component, Y_j, with the ith original variable, X_i.

Suppose $\mathbf{x}$ is standardized to have zero mean vector and unit variance for each X_i. Now, $\mathbf{y} = \mathbf{A}'\mathbf{x}$, or $\mathbf{A}\mathbf{y} = \mathbf{x}$. The covariance between Y_j and X_i is given by

$$
\operatorname{cov}(Y_j, X_i) = \operatorname{cov}\left(Y_j, \sum_{k=1}^{p} a_{ik} Y_k\right)
$$

$$
= \sum_{k=1}^{p} a_{ik} \operatorname{cov}(Y_j, Y_k)
$$

$$
= a_{ij} \operatorname{var}(Y_j) \quad \text{(since the } Y_j\text{'s are uncorrelated)}
$$

$$
= a_{ij} \lambda_j.
$$

Then, since the standard deviation of Y_j is $\lambda_j^{1/2}$,

$$
\operatorname{corr}(Y_j, X_i) = a_{ij} \lambda_j / \lambda_j^{1/2} = a_{ij} \lambda_j^{1/2}.
$$

Putting all the correlations together into a matrix,

$$
\operatorname{corr}(\mathbf{y}, \mathbf{x}) = \mathbf{A} \mathbf{\Lambda}^{1/2},
$$

where $\mathbf{\Lambda}^{1/2} = \operatorname{diag}(\sqrt{\lambda_1}, \ldots, \sqrt{\lambda_p})$.

Table 5.2 Component correlations for the bodyfat data

PC:	PC1	PC2	PC3	PC4	PC5	PC6	PC7	PC8	PC9	PC10
neck	0.87	0.00	0.21	0.24	0.03	0.15	0.33	−0.13	−0.02	−0.01
chest	0.90	0.23	0.05	0.17	−0.24	−0.04	−0.11	−0.05	0.15	−0.11
abdomen	0.89	0.34	−0.05	0.15	−0.17	−0.08	−0.06	0.02	−0.08	0.17
hip	0.92	0.22	−0.17	−0.08	0.03	−0.04	0.03	0.14	−0.15	−0.14
thigh	0.88	0.16	−0.15	−0.29	0.14	0.06	0.13	0.16	0.15	0.06
knee	0.87	−0.02	−0.22	−0.09	0.24	−0.23	−0.05	−0.25	0.00	0.00
ankle	0.65	−0.53	−0.48	−0.02	−0.23	0.09	0.03	−0.01	−0.01	0.01
biceps	0.85	−0.02	0.21	−0.21	0.05	0.35	−0.21	−0.07	−0.04	0.01
forearm	0.72	−0.31	0.48	−0.28	−0.14	−0.23	0.04	0.04	−0.02	0.01
wrist	0.79	−0.32	0.12	0.40	0.23	−0.04	−0.12	0.16	0.02	0.01

Table 5.2 shows the component correlations for the bodyfat data, where the sample correlation matrix has been used for the principal components.

Not surprisingly, all ten variables are highly correlated with the first principal component. Looking at all positive and negative correlations of 0.3 or more, we see that the neck circumference is moderately correlated with PC7, abdomen with PC2, biceps with PC6, forearm with PC2 and PC3, and perhaps surprisingly, ankle with PC2 and PC3, and wrist with PC2 and PC4.

5.5 Further comments

In this section we give brief details of some further aspects of principal components analysis.

5.5.1 Reification – interpretation of principal components

Once principal components have been found, we might try to interpret them, giving them names, for instance 'size' for the first principal component of the bodyfat data. This is known as reification. Sometimes the PCs have clear interpretations in terms of the objects measured and variables recorded, but sometimes such interpretations appear artificially contrived. When there are only a few variables, then the constraint of orthogonality will restrict the number of possible coefficient patterns among the principal components. When PCA is part of factor analysis (see Chapter 15), the reification might become simpler.

5.5.2 Geometric interpretation

Geometrically, principal components analysis can be viewed as an orthogonal rotation, and possibly a reflection, of the (orthogonal) set of axes representing the original variables. The axes denoted by $(X_1, \ldots, X_p)$ are rotated to give a new set of axes $(Y_1, \ldots, Y_p)$, where the projection of the data points onto Y_1 gives the maximum variance possible for the projected points, and the projection onto Y_2 similarly gives the maximum variance, but of course with

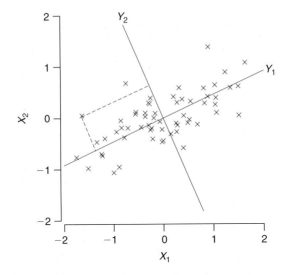

Figure 5.6 Orthogonal rotation of X_1 and X_2 to PC1 and PC2

the constraint that Y_2 is orthogonal to Y_1, and so forth. Figure 5.6 shows a set of orthogonal axes, X_1 and X_2, together with data plotted for the two variables. These axes have then been orthogonally rotated to Y_1 and Y_2, such that Y_1 has the largest possible variance of the points projected onto Y_1, and Y_2 has the second largest variance. (Of course, since we have only two variables, Y_2 is determined, once Y_1 has been found, because of the orthogonality constraint.) The rotation from (X_1, X_2) to (Y_1, Y_2) is given by

$$Y_1 = 0.91X_1 + 0.42X_2$$
$$Y_1 = -0.42X_1 + 0.91X_2.$$

The rotation matrix is

$$\begin{bmatrix} 0.91 & 0.42 \\ -0.42 & 0.91 \end{bmatrix},$$

and this is the same as matrix $\mathbf{A}'$ for our principal components.

This is the situation in two dimensions. Now consider a similar situation in three dimensions. Let a misshaped rugby ball, or American football, which has been squashed to make its cross-section elliptical and not circular, represent the cloud of data points. The first principal component is given by an axis drawn from end to end of the ball, with origin at the centre. The second and third principal components are orthogonal axes passing through the centre, and lying in the cross-sectional plane. In like manner, in p dimensions, we have the principal components as the principal axes of a p-dimensional ellipsoid.

5.5.3 Inference for PCA

Denote $\hat{\lambda}_j$ and $\hat{\mathbf{a}}_j$ as the jth eigenvalue and eigenvector of $\mathbf{S}$. These are the sample principal components which estimate λ_j and $\mathbf{a}_j$, the eigenvalues and eigenvectors of $\mathbf{\Sigma}$, the population principal components. Let $\mathbf{\Lambda} = \mathrm{diag}(\lambda_1, \ldots, \lambda_p)$, and we assume that there are no repeated eigenvalues. If the data, $\mathbf{x}_1, \ldots, \mathbf{x}_n$, are a random sample from a multivariate normal distribution, $N_p(\boldsymbol{\mu}, \boldsymbol{\Sigma})$, with $\boldsymbol{\Sigma}$ positive definite, then it can be shown that

$$\sqrt{n}(\hat{\boldsymbol{\lambda}} - \boldsymbol{\lambda}) \approx N_p(\mathbf{0}, 2\mathbf{\Lambda}^2).$$

Let $\mathbf{E}_j = \lambda_j \sum_{i \neq j} \frac{\lambda_i}{(\lambda_i - \lambda_j)^2} \mathbf{a}_i \mathbf{a}_i'$, and then it can also be shown that

$$\sqrt{n}(\hat{\mathbf{a}}_j - \mathbf{a}_j) \approx N_p(\mathbf{0}, \mathbf{E}_j).$$

Also, each $\hat{\lambda}_j$ is distributed independently of the elements of $\hat{\mathbf{a}}_j$.

So, for reasonably large sample sizes, the $\hat{\lambda}_j$'s are independently distributed as $N(\lambda_j, 2\lambda_j^2/n)$, and hence we can find an approximate confidence interval for λ_j,

$$\frac{\hat{\lambda}_j}{(1 + z_{\alpha/2}\sqrt{2/n})} \leq \lambda_j \leq \frac{\hat{\lambda}_j}{(1 - z_{\alpha/2}\sqrt{2/n})},$$

or alternatively test hypotheses about the value. Other hypothesis tests can be carried out, such as testing for the equality of several of the eigenvectors of $\boldsymbol{\Sigma}$. Further details can be found in Jolliffe (2002), Mardia et al. (1979), or Jackson (1991).

5.6 Golf

The scores for thirty-six golfers for the first round of the 130th Open Championship held at the Lytham St Annes golf course were subjected to PCA. For each of the eighteen holes, each golfer records the number of shots that they took to get the golf ball from the tee, to the bottom of the hole on the green. Each hole has a designated 'par', which is the guideline number of shots needed to complete that hole. Table 5.3 gives the mean number of shots taken for each hole, together with the standard deviation and par for the hole.

We see that the mean shots for holes 1, 3, 5, 12, 14, 15, and 17 are greater than their corresponding par values, suggesting these are the more difficult holes to play. The standard deviations range from 0.50 for hole 18, to 0.81 for hole 14.

Table 5.3 Mean, standard deviation and par for the 18 holes

Hole	1	2	3	4	5	6	7	8	9
$\bar{x}$	3.14	3.83	4.22	4.00	3.14	4.36	4.56	3.86	2.89
s	0.59	0.56	0.54	0.59	0.64	0.72	0.74	0.76	0.57
par	3	4	4	4	3	5	5	4	3

Hole	10	11	12	13	14	15	16	17	18
$\bar{x}$	3.92	4.61	3.33	3.83	4.25	4.36	3.86	4.17	3.92
s	0.50	0.60	0.63	0.61	0.81	0.59	0.59	0.61	0.50
par	4	5	3	4	4	4	4	4	4

Table 5.4 Correlation matrix ($\times 100$) for the golf data

	h1	h2	h3	h4	h5	h6	h7	h8	h9	h10	h11	h12	h13	h14	h15	h16	h17	h18
h1	.	07	08	-16	02	21	08	-02	05	14	-01	11	06	-19	-14	30	-14	-15
h2	.	.	-07	-17	-33	23	03	09	-15	-15	06	-33	34	-10	01	10	33	-25
h3	.	.	.	00	16	-29	-25	01	08	-04	-17	28	03	20	-16	01	-29	-04
h4	.	.	.	.	00	-27	-26	32	-17	20	-08	-15	-32	06	00	-08	00	00
h5	.	.	.	.	.	-05	-05	-37	-27	-14	-31	-04	-09	38	-13	20	-06	-05
h6	.	.	.	.	.	.	06	05	-17	-07	41	-03	08	-17	07	-21	18	01
h7	.	.	.	.	.	.	.	-26	-05	-33	-13	01	09	-20	-03	-01	10	-18
h8	.	.	.	.	.	.	.	.	03	27	38	-08	-23	01	-02	-23	-07	05
h9	.	.	.	.	.	.	.	.	.	27	-04	34	-05	-37	28	-05	-03	37
h10	.	.	.	.	.	.	.	.	.	.	-01	00	-14	-16	01	25	14	32
h11	.	.	.	.	.	.	.	.	.	.	.	-03	06	-04	-09	-23	-06	08
h12	.	.	.	.	.	.	.	.	.	.	.	.	07	-22	-17	-33	-29	-09
h13	.	.	.	.	.	.	.	.	.	.	.	.	.	-21	-23	02	23	-23
h14	.	.	.	.	.	.	.	.	.	.	.	.	.	.	-19	01	03	05
h15	.	.	.	.	.	.	.	.	.	.	.	.	.	.	.	-10	15	20
h16	.	.	.	.	.	.	.	.	.	.	.	.	.	.	.	.	14	15
h17	.	.	.	.	.	.	.	.	.	.	.	.	.	.	.	.	.	-14
h18	.	.	.	.	.	.	.	.	.	.	.	.	.	.	.	.	.	.

The correlation matrix for the shots taken is given in Table 5.4, which gives rise to the first nine principal components:

PC:	PC1	PC2	PC3	PC4	PC5	PC6	PC7	PC8	PC9
E'value	1.06	1.00	0.82	0.66	0.55	0.52	0.45	0.36	0.31
Propn.	0.15	0.14	0.12	0.09	0.08	0.07	0.06	0.05	0.04
Cum.	0.15	0.29	0.41	0.50	0.58	0.66	0.72	0.77	0.82
h1	-0.14	0.09	-0.06	-0.10	-0.45	-0.18	0.45	-0.07	-0.22
h2	-0.19	0.04	0.32	0.11	-0.23	0.26	-0.15	-0.10	-0.39
h3	0.19	-0.02	-0.21	-0.20	-0.24	0.12	-0.16	-0.09	-0.35
h4	0.19	-0.27	-0.01	0.18	0.09	0.19	0.22	0.52	0.06
h5	0.37	0.28	0.02	-0.11	-0.03	-0.32	0.08	0.31	0.04
h6	-0.36	-0.01	0.40	-0.29	0.00	-0.53	0.02	0.19	-0.04
h7	-0.24	0.42	0.01	0.00	0.44	0.24	0.45	-0.31	0.15
h8	-0.06	-0.65	0.13	-0.06	-0.04	0.23	0.26	-0.13	-0.02
h9	-0.17	-0.10	-0.40	0.17	0.01	-0.13	-0.25	-0.28	0.06
h10	-0.01	-0.20	-0.11	0.24	-0.26	-0.15	0.05	0.05	0.35
h11	-0.19	-0.27	0.23	-0.28	-0.01	-0.11	-0.03	-0.21	0.24
h12	-0.11	0.01	-0.45	-0.43	0.01	0.02	-0.14	0.11	0.17
h13	-0.20	0.23	0.09	-0.11	-0.32	0.29	-0.38	0.07	0.18
h14	0.62	0.00	0.36	-0.19	0.09	-0.03	-0.15	-0.39	0.05
h15	-0.12	-0.08	-0.04	0.35	0.33	-0.28	-0.20	0.03	-0.49
h16	0.11	0.18	0.01	0.38	-0.44	-0.13	0.23	-0.21	0.08
h17	-0.11	0.10	0.33	0.35	-0.02	0.07	-0.67	0.15	0.34
h18	0.05	-0.14	-0.11	0.16	0.05	-0.35	-0.10	-0.32	0.20

The first principal component accounts for only 15% of the total variation and, perhaps surprisingly, does not measure overall score for the golfers. It is dominated by the score for hole 14, the one with the largest standard deviation. This score and the score for hole 5 is contrasted with the score for hole 6. The second principal component accounts for 14% of the total variation, and contrasts the score for hole 8 with that for hole 7. The third principal component accounts for 12% of the total variation, and contrasts the scores for holes 2, 6,

14 and 17, with those for holes 9 and 12. We would need seven PCs to explain 70% of the total variation in the data.

Figure 5.7 shows the component scores for the first two principal components plotted against each other using the initials of the individual golfers. Figure 5.8 shows the component scores for the first and third principal components. A few of the golfers stand out from the rest (NF, FC, PM and BF): NF has the lowest score on PC2; FC has the highest score on PC1 and on PC3; PM has a high score on PC1 and a low score on PC3; BF has a low score on PC3.

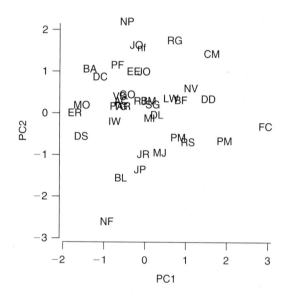

Figure 5.7 PC2 versus PC1 for the golfers

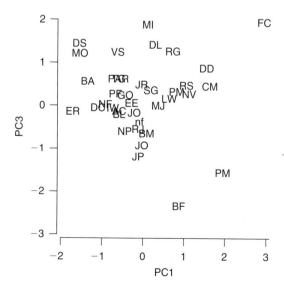

Figure 5.8 PC3 versus PC1 for the golfers

The same PCA was carried out on the scores for round two of the tournament, and Figure 5.9 shows the plot of the component scores for the first two principal components, where this time golfers MJ, JR, MI, NF and FC stand out.

The first principal component for round two is given by

$$PC1 = (-0.06, -0.22, 0.03, 0.00, 0.05, 0.04, -0.06, -0.13, -0.30,$$
$$-0.29, -0.72, -0.02, -0.11, 0.14, -0.03, 0.18, 0.25, -0.33)'.$$

This first principal component explains 16% of the total variation in the data, similar to the amount explained by PC1 for the first round. Although the coefficients are different for PC1 for the two rounds, a similar pattern of positive and negative coefficients occur. (Note, it may have been necessary to have reversed the signs for one of the PC1s in order to achieve this alignment due to the arbitrary sign of the eigenvectors.) The holes with positive coefficients for both rounds are 3, 4, 5, 14 and 16 (group 1). The holes with negative coefficients for both rounds are 1, 2, 7, 8, 9, 10, 11, 12, 13 and 15 (group 2). Placing the remaining holes (6, 7 and 18) into one or other of the two groups according to the size of their coefficients for the two PC1's, gives the two contrasting groups of holes

Group 1:   3, 4, 5, 14, 16, 17

Group 2:   1, 2, 6, 7, 8, 9, 10, 11, 12, 13, 15, 18.

The grouping of the holes suggests that the nature of the holes in the two groups could be different. Golfers who do well on one group do less well on the other, and vice versa. This conclusion has some support from the fact that the correlation of the total score for the holes in group 1, with the total score for the holes in group 2, is −0.18.

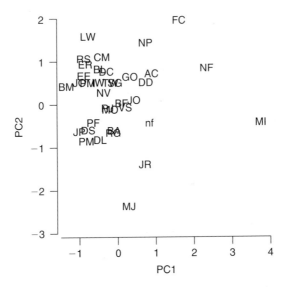

Figure 5.9 PC2 versus PC1 for round 2

5.7 Exercises

1. Fourteen students take three exams for a university mathematics module. The marks are given below.

student	1	2	3	4	5	6	7	8	9	10	11	12	13	14
exam1	46	31	43	43	47	51	40	48	27	50	31	49	44	43
exam2	55	26	55	42	44	52	40	54	26	52	30	44	54	42
exam3	67	30	81	61	52	66	45	54	9	77	5	48	77	56

Show that the covariance matrix is given by

$$\begin{pmatrix} 57.5 & 69.0 & 144.6 \\ 69.0 & 110.6 & 222.8 \\ 144.6 & 222.8 & 555.4 \end{pmatrix}.$$

(a) Find the principal components based on the covariance matrix, and the variation explained by each. Interpret the principal components.
(b) Find the component scores and plot the scores for PC1 against those for PC2.
(c) Find the component correlations.
(d) Find the correlation matrix and the principal components based on the correlation matrix.

2. Let (X_1, X_2) have correlation matrix given by

$$\Sigma = \begin{pmatrix} 1 & \rho \\ \rho & 1 \end{pmatrix}.$$

Find the principal components and the amount of variation explained by each. What happens when $\rho = 0$? What happens when $\rho = 1$?

3. Let (X_1, X_2, X_3) have correlation matrix given by

$$\Sigma = \begin{pmatrix} 1 & \rho & \rho \\ \rho & 1 & \rho \\ \rho & \rho & 1 \end{pmatrix},$$

where $\rho > 0$.

(a) Show that the eigenvectors of Σ are $1 + 2\rho, 1 - \rho, 1 - \rho$, and thus PC2 and PC3 are not unique.
(b) Show that the first principal component is given by

$$Y_1 = \frac{(X_1 + X_2 + X_3)}{\sqrt{3}}.$$

(c) Let the eigenvectors of the second two eigenvalues be (a_1, a_2, a_3) and (b_1, b_2, b_3). Show that these satisfy the following system of equations

$$a_1^2 + a_2^2 + a_3^2 = 1$$
$$b_1^2 + b_2^2 + b_3^2 = 1$$
$$a_1 + a_2 + a_3 = 0$$
$$b_1 + b_2 + b_3 = 0$$
$$a_1 b_1 + a_2 b_2 + a_3 b_3 = 0.$$

(d) Hence show that PC2 and PC3 are given by

$$Y_2 = \alpha X_1 + \frac{1}{2}(-\alpha \pm \beta)X_2 + \frac{1}{2}(-\alpha \mp \beta)X_3$$

$$Y_3 = \frac{1}{\sqrt{3}}\beta X_1 + \frac{1}{2\sqrt{3}}(\mp \alpha - \beta)X_2 + \frac{1}{2\sqrt{3}}(\pm \alpha - \beta)X_2,$$

where $\beta = (2 - 3\alpha^2)^{1/2}$, and where $-\sqrt{2/3} \le \alpha \le \sqrt{2/3}$.

(e) Investigate the special case where any one of the elements, a_1, a_2, a_3, b_1, b_2 or b_3 is put equal to zero.

4. Download the chemical composition of pottery data from Chania and analyse them using principal components analysis on the correlation matrix. Show that the first three principal components account for 80% of the variation. Interpret them. Plot component scores and look for patterns in the data. Find confidence intervals for the first two eigenvalues of the correlation matrix, and hence for the variance of PC1 and PC2.

6

Biplots

Most graphical displays of multivariate data concern either the observations (individuals or objects) or the variables. In the previous chapter on principal components analysis, a plot of component scores is a graphical display for the observations, whilst a plot of coefficients for the first principal component against those for the second is a graphical display for the variables. Biplots illustrate both the observations and the variables in the same plot, and hence the 'bi' in the name. Biplots were introduced by Gabriel (1971). Gower and Hand (1996) is an authoritative monograph on the subject.

Biplots can be constructed for continuous variables and categorical variables. The aim is to find a space in which points represent the observations, and then for continuous variables, axes are placed in the space, each axis representing one of the variables. The axes are usually linear, but can be non-linear. For a categorical variable, a simplex of points is overlaid in the space, with each point in the simplex representing one of the categories for the variable. The original and most popular biplots are those based on principal components analysis, with points representing the observations and drawn vectors representing the variables.

6.1 The classic biplot

The classic biplot is for continuous variables. It represents the rows and columns of an $n \times p$ data matrix, $\mathbf{X}$, as vectors in a two-dimensional space. Let the SVD of $\mathbf{X}$ be given by

$$\mathbf{X} = \mathbf{U}\boldsymbol{\Gamma}\mathbf{V}'.$$

We now approximate $\mathbf{X}$ using the first two singular values and associated singular vectors,

$$\mathbf{X} \approx \mathbf{U}_2\boldsymbol{\Gamma}_2\mathbf{V}_2', \tag{6.1}$$

where the suffix 2 means keep the first two singular values and vectors and disregard the rest. Now write (6.1) as

$$\mathbf{X} \approx (\mathbf{U}_2\boldsymbol{\Gamma}_2^{\alpha})(\mathbf{V}_2\boldsymbol{\Gamma}_2^{1-\alpha})',$$

where α is a chosen constant with $0 \leq \alpha \leq 1$. Different values of α give rise to different biplots.

The $n \times 2$ matrix $\mathbf{U}_2\boldsymbol{\Gamma}_2^{\alpha}$ consists of n row vectors representing the rows, and hence the observations, of the matrix $\mathbf{X}$. The $p \times 2$ matrix $(\mathbf{V}_2\boldsymbol{\Lambda}_2^{1-\alpha})$ consists of p column vectors

Table 6.1 Soil data

Sample no.	Sand content (%)	Silt content (%)	Clay content (%)	Organic matter (%)	pH
1	77.3	13.0	9.7	1.5	6.4
2	82.5	10.0	7.5	1.5	6.5
3	66.9	20.6	12.5	2.3	7.0
4	47.2	33.8	19.0	2.8	5.8
5	65.3	20.5	14.2	1.9	6.9
6	83.3	10.0	6.7	2.2	7.0
7	81.6	12.7	5.7	2.9	6.7
8	47.8	36.5	15.7	2.3	7.2
9	48.6	37.1	14.3	2.1	7.2
10	61.6	25.5	12.9	1.9	7.3
11	58.6	26.5	14.9	2.4	6.7
12	69.3	22.3	8.4	4.0	7.0
13	61.8	30.8	7.4	2.7	6.4
14	67.7	25.3	7.0	4.8	7.3
15	57.2	31.2	11.6	2.4	6.5
16	67.2	22.7	10.1	3.3	6.2
17	59.2	31.2	9.6	2.4	6.0
18	80.2	13.2	6.6	2.0	5.8
19	82.2	11.1	6.7	2.2	7.2
20	69.7	20.7	9.6	3.1	5.9

representing the columns of **X**. These vectors are then plotted in a two-dimensional space, one set representing the observations, and the other, the variables. For convenience of plotting, biplots are usually constructed in two dimensions. However by using three or more singular values and corresponding singular vectors in the approximation of **X**, higher dimensional plots can be obtained.

Example

Table 6.1 shows the content of twenty soil samples. The variables measured are *sand content* (x_1), *silt content* (x_2), *clay content* (x_3), *organic matter* (x_4), and *pH* (x_5). The data are displayed and analysed in Kendall (1975) using principal components analysis.

Since the first three variables are percentage data, they sum to 100, essentially making one of the variables redundant. Thus we exclude x_1 from the analysis.

First the data are mean corrected. These are shown below with sand content excluded.

Sample no.	Silt content	Clay content	Organic matter	pH
1	−9.74	−0.81	−1.04	−0.25
2	−12.74	−3.01	−1.04	−0.15
3	−2.14	2.00	−0.24	0.35
4	11.07	8.50	0.27	−0.85
5	−2.24	3.70	−0.64	0.25
6	−12.74	−3.81	−0.34	0.35

(Contd)

Sample no.	Silt content	Clay content	Organic matter	pH
7	−10.04	−4.81	0.37	0.05
8	13.77	5.20	−0.24	0.55
9	14.37	3.80	−0.44	0.55
10	2.77	2.40	−0.64	0.65
11	3.77	4.40	−0.14	0.05
12	−0.44	−2.11	1.47	0.35
13	8.07	−3.11	0.17	−0.25
14	2.57	−3.51	2.27	0.65
15	8.47	1.10	−0.14	−0.15
16	−0.04	−0.41	0.77	−0.45
17	8.47	−0.91	−0.14	−0.65
18	−9.54	−3.91	−0.54	−0.85
19	−11.64	−3.81	−0.34	0.55
20	−2.04	−0.91	0.57	−0.75

The SVD of **X** gives the **U**, **Γ** and **V** matrices:

$$
U = \begin{bmatrix}
-0.24 & -0.19 & 0.20 & -0.07 \\
-0.32 & -0.09 & 0.24 & -0.02 \\
-0.04 & -0.22 & -0.08 & 0.15 \\
0.32 & -0.41 & -0.23 & -0.43 \\
-0.03 & -0.36 & -0.02 & 0.11 \\
-0.33 & -0.01 & 0.02 & 0.17 \\
-0.27 & 0.15 & -0.09 & 0.01 \\
0.36 & -0.08 & 0.08 & 0.26 \\
0.37 & 0.04 & 0.22 & 0.28 \\
0.08 & -0.13 & 0.12 & 0.32 \\
0.12 & -0.27 & -0.09 & 0.01 \\
-0.02 & 0.18 & -0.40 & 0.08 \\
0.17 & 0.46 & 0.26 & -0.07 \\
0.04 & 0.38 & -0.55 & 0.19 \\
0.21 & 0.12 & 0.17 & -0.04 \\
0.00 & 0.04 & -0.21 & -0.24 \\
0.19 & 0.28 & 0.28 & -0.25 \\
-0.25 & 0.07 & 0.21 & -0.34 \\
-0.30 & 0.02 & 0.03 & 0.26 \\
-0.05 & 0.03 & -0.15 & -0.37
\end{bmatrix}
$$

$$
\Gamma = \begin{bmatrix}
40.57 & 0 & 0 & 0 \\
0 & 11.61 & 0 & 0 \\
0 & 0 & 2.99 & 0 \\
0 & 0 & 0 & 2.22
\end{bmatrix}
$$

$$
V = \begin{bmatrix}
0.96 & 0.29 & 0.06 & 0.01 \\
0.29 & -0.95 & -0.14 & -0.02 \\
0.01 & 0.15 & -0.98 & -0.14 \\
0.00 & 0.00 & -0.14 & 0.99
\end{bmatrix}.
$$

First, we choose $\alpha = 1$. The matrix $\mathbf{U\Gamma}$ is

$$\mathbf{U\Gamma} = \begin{bmatrix} -9.56 & -2.20 & 0.59 & -0.15 \\ -13.07 & -0.99 & 0.71 & -0.03 \\ -1.46 & -2.54 & -0.23 & 0.33 \\ 13.07 & -4.80 & -0.69 & -0.96 \\ -1.06 & -4.23 & -0.07 & 0.25 \\ -13.29 & -0.12 & 0.07 & 0.38 \\ -11.00 & 1.71 & -0.28 & 0.02 \\ 14.68 & -0.98 & 0.23 & 0.57 \\ 14.84 & 0.48 & 0.66 & 0.63 \\ 3.34 & -1.56 & 0.36 & 0.70 \\ 4.89 & -3.09 & -0.27 & 0.01 \\ -1.01 & 2.09 & -1.21 & 0.18 \\ 6.80 & 5.28 & 0.79 & -0.16 \\ 1.46 & 4.40 & -1.66 & 0.42 \\ 8.41 & 1.38 & 0.50 & -0.10 \\ -0.14 & 0.49 & -0.63 & -0.54 \\ 7.82 & 3.27 & 0.85 & -0.56 \\ -10.27 & 0.86 & 0.63 & -0.76 \\ -12.24 & 0.19 & 0.10 & 0.59 \\ -2.20 & 0.35 & -0.44 & -0.82 \end{bmatrix}.$$

The first two columns of $\mathbf{U\Gamma}$ give the coordinates of the points representing the observations (sample no.). These are plotted in Figure 6.1. The vectors in the plot, representing the variables, have been drawn using the first two columns of $\mathbf{V}$. The usual way of drawing the

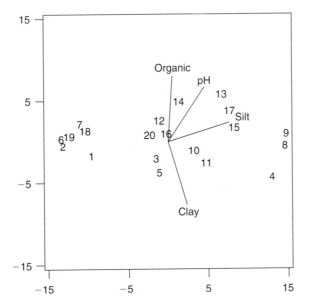

Figure 6.1 Biplot for the soil data

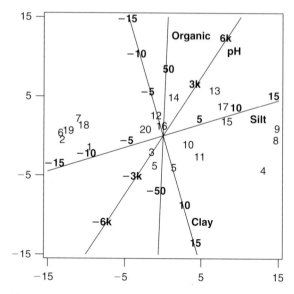

Figure 6.2 Biplot with axes

vectors representing the variables is to make the length of them equal to a multiple of their standard deviation. It is the direction that is more important. Now, because the standard deviations for organic matter and pH are very small compared to those of silt content and clay content, and would not be easily seen in the plot, the vectors in Figure 6.1 have been plotted with constant length. However, Gower and Hand (1996) recommend drawing these as true axes with scales marked on appropriately. These can be seen in Figure 6.2.

From the figures, there is a group of samples, {1, 2, 6, 7, 18, 19}, to the left of the plot, and these have low silt content, whereas the group to the right {4, 8, 9}, have high silt content. The projection of an individual point onto the four axes representing the variables gives approximately the scores for these variables, for that point.

6.1.1 Principal components biplot

Again, let $\mathbf{X}$ be mean corrected. Then the sample covariance matrix, $\mathbf{S}$, is given by $(n-1)\mathbf{S} = \mathbf{X}'\mathbf{X}$. Replace $\mathbf{X}$ by its SVD, and hence

$$(n-1)\mathbf{S} = \mathbf{V}\mathbf{\Gamma}\mathbf{U}'\mathbf{U}\mathbf{\Gamma}\mathbf{V}' = \mathbf{V}\mathbf{\Gamma}^2\mathbf{V}'.$$

This is now the spectral decomposition of $\mathbf{S}$, giving the principal components as the columns of $\mathbf{V}$. The components scores are given by $\mathbf{X}\mathbf{V} = \mathbf{U}\mathbf{\Gamma}$. Thus the biplot where $\alpha = 1$ corresponds to a principal components analysis, and hence the term principal components biplot.

The points in Figures 6.1 and 6.2 are thus a plot of the component scores for the first two principal components. The angle of the vectors representing the variables, to the x-axis, i.e. the first PC, indicate the contribution of the variable to the first PC. A narrow angle indicates that the variable plays a major role in the PC. So silt content is important in the first PC.

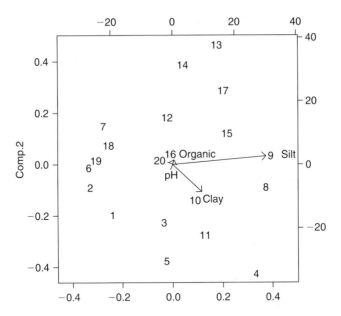

Figure 6.3 *J'K* biplot

Similarly, clay content is important in the second PC. Note that since we scaled the vectors to have the same length, we cannot infer the actual component loadings from the plot.

6.1.2 *J'K* biplot

When $\alpha = 0$, the biplot is called a *J'K* biplot. Figure 6.3 shows a *J'K* biplot for the soil data, generated using S-Plus. The coordinates of the points are given by the first two columns of **U**. The left and lower axes relate to these coordinates. The vectors representing the variables are given by the first two columns of $\mathbf{\Gamma_2 V}$. The right and upper axes relate to these. Again, because the variances of clay content and silt content dominate those for organic matter and pH, the vectors for organic matter and pH are very short. The *J'K* biplot places more emphasis on the variables than the principal components biplot, with the lengths of the vectors for the variables being approximately equal to their standard deviations.

6.2 Non-linear axes

First, we briefly describe how categorical variables can be illustrated within a biplot, using the notion of a *pseudosample*. Suppose a configuration of points has been constructed that represent the observations. This could be achieved using multidimensional scaling (see Chapter 9) and Gower's general dissimilarity coefficient, for instance. Now let the observation on the *i*th variable be replaced by the value y for all the n observations. These

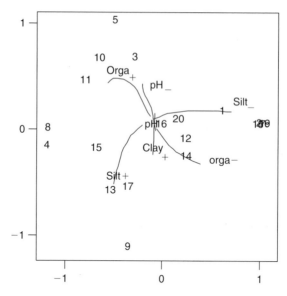

Figure 6.4 Non-linear axes

are then n pseudosamples for the ith variable,

$$\mathbf{x}_r(\gamma) = (x_{r,1}, \ldots, x_{r,i-1}, \gamma, x_{r,i+1}, \ldots, x_{r,p})' \quad (r = 1, \ldots, n).$$

The values that γ can take are the category levels for the ith variable. For a particular value of γ, each pseudosample is superimposed upon the configuration in some appropriate manner. (See Gower and Hand (1996) for a detailed description of how this can be done.) Then the centroid of these added points is a point that represents this particular category of ith variable. This is repeated for all the levels of the ith variable, resulting in a simplex of points within the space, which represent the ith variable. The process is repeated for all the categorical variables.

For a continuous variable, the same process is followed, but γ can now vary over the range the variable. Figure 6.4 illustrates this for the soil data. This time the four variables have not only been mean corrected, but also standardized so that their standard deviations are unity. The coordinates of the points are given by the scores for the first two principal components. The four variables have then been added to the plot using the process of pseudosamples as described above. The axes have become very non-linear, especially for silt content. The positive end of the axis bends to the centre of the group of highest scoring points for silt content (4, 8, 9, 4, 13, 15, 17), while the negative end bends to the tight group of low scoring points for silt content (2, 6, 7, 18, 19).

6.3 Exercises

1. Find a principal components biplot for the soil data, having standardized the variables to have standard deviation equal to unity.

2. Find a suitable biplot for the Chania data described in Chapter 4.

3. Cox and Cox (2000) give a table of scores for composition, drawing, colour and expression, for ten Renaissance painters. The scores are

	Composition	Drawing	Colour	Expression
Del Sarto	12	16	9	8
Del Piombo	8	13	16	7
Da Udine	10	8	16	3
Guilio Romano	15	16	4	14
Da Vinci	15	16	4	14
Michelangelo	8	17	4	8
Fr. Penni	0	15	8	0
Perino del Vaga	15	16	7	6
Perugino	4	12	10	4
Raphael	17	18	12	18

These scores are a subset of those made by Roger de Piles in the seventeenth century. The data can be found in Davenport and Studdert-Kennedy (1972).

Find a suitable biplot for these data.

7

Correspondence analysis

Correspondence analysis is a method for graphically displaying both the rows and columns of a two-way contingency table. It can also be used on any data matrix that has non-negative elements. The method has been discovered and rediscovered several times and has other names, such as reciprocal averaging and dual scaling. Greenacre (1993) gives a thorough introduction to the subject.

7.1 Distance for contingency tables

First, we define distances between the rows of a contingency table, which also equally apply to the columns of the table. Table 7.1 shows a contingency table of 298 women in corporate sales, classified by position and type of dress (Position: P1 – *Top*, P2 – *Above Average*, P3 – *Average or Below Average*, P4 – *Failing*; Dress: Prof – *Professionally Dressed*, Appr – *Appropriately Dressed*, Fash – *Fashionably Dressed*, Poor – *Poorly Dressed*). The data are used on a statistics course at Napier University (http://www.maths.napier.ac.uk/~jeff/courses/mmmult.html), and originate from an article in the *Wall Street Journal*, 1 September 1987.

Treating the rows as four points in a four-dimensional Euclidean space, each column defining one of the dimensions, would not be a useful graphical way of representing the data. The Euclidean distance between rows is affected by the row totals and also the overall total. For example, the distance between P1 and P2 is

$$\{(12 - 56)^2 + (12 - 39)^2 + (7 - 18)^2 + (1 - 16)^2\}^{1/2} = 54.9.$$

This value is greatly affected by the number of P1 and P2 women. If the number of women in one category was doubled, but the proportions in the columns stayed the same, then the

Table 7.1 Corporate sales data

	Prof	Appr	Fash	Poor	Total
P1	12	12	7	1	32
P2	56	39	18	16	129
P3	20	42	32	26	120
P4	4	2	2	9	17
Total	92	95	59	52	298

Table 7.2 Standardized data

	Prof	Appr	Fash	Poor	Total
P1	0.040	0.040	0.023	0.003	0.107
P2	0.188	0.131	0.060	0.054	0.433
P3	0.067	0.141	0.107	0.087	0.403
P4	0.013	0.007	0.007	0.030	0.057
Total	0.308	0.319	0.197	0.174	1.0

Table 7.3 Row profiles

	Prof	Appr	Fash	Poor	Total
P1	0.375	0.375	0.219	0.031	1.0
P2	0.434	0.302	0.140	0.124	1.0
P3	0.167	0.350	0.267	0.217	1.0
P4	0.235	0.118	0.118	0.529	1.0

Table 7.4 Column profiles

	Prof	Appr	Fash	Poor
P1	0.130	0.126	0.119	0.019
P2	0.609	0.410	0.305	0.308
P3	0.217	0.442	0.542	0.500
P4	0.043	0.021	0.034	0.173
Total	1.0	1.0	1.0	1.0

distance between the rows would change. This is not particularly desirable. To overcome this, first the data are standardized dividing each cell frequency by the overall total. Table 7.2 gives the standardized data. (Note that rounding errors have given some discrepancies.)

Now *row profiles* are produced, where each cell frequency in a row is divided by the row total, thus standardizing over rows. Table 7.3 shows the row profiles for the corporate sales data. It is now easier to compare the four groups of women.

Similarly *column profiles* can be found, standardizing over the columns, as shown in Table 7.4. From the column profiles, comparisons between the types of dress can be made.

Let the data matrix be $\mathbf{X}$, which has been standardized so that $\sum_r \sum_i x_{ri} = 1$, i.e. each element of the original data matrix is divided by the total sum of all the elements. Let $\mathbf{X}$ have n rows and p columns. Let the rth row total be R_r. Let the ith column total be C_i. Let the row totals be placed in the diagonal matrix $\mathbf{D}_R$, and so $\mathbf{D}_R = \mathrm{diag}(R_1, \ldots, R_n)$. Let the column totals be placed in a diagonal matrix $\mathbf{D}_C$, where $\mathbf{D}_C = \mathrm{diag}(C_1, \ldots, C_p)$. The matrix of row profiles is then given by $\mathbf{D}_R^{-1}\mathbf{X}$, and the matrix of column profiles by $\mathbf{D}_C^{-1}\mathbf{X}$.

Distances between rows in $\mathbf{X}$ are defined on the row profiles, with the distance between row r and row s being given by

$$d_{rs}^2 = \sum_{i=1}^{p} \frac{1}{C_i} \left(\frac{x_{ri}}{R_r} - \frac{x_{si}}{R_s} \right)^2. \tag{7.1}$$

This is a weighted Euclidean distance and is called the χ^2 *distance* between rows.

The χ^2 distance between columns i and j is similarly defined as

$$d_{ij}^2 = \sum_{r=1}^{n} \frac{1}{R_r} \left(\frac{x_{ri}}{C_i} - \frac{x_{rj}}{C_j} \right)^2.$$

Note that distance between a row and a column is not defined.

For our corporate sales data, the standardized matrix, $\mathbf{X}$, is given by the cell entries in Table 7.2, and the row and column profile matrices, $\mathbf{D}_R^{-1}\mathbf{X}$ and $\mathbf{D}_C^{-1}$, are given by the cell entries of Tables 7.3 and 7.4, respectively.

The matrix of χ^2 distances between rows is

$$\begin{bmatrix} 0.0 & 0.330 & 0.593 & 1.320 \\ 0.330 & 0.0 & 0.608 & 1.086 \\ 0.593 & 0.608 & 0.0 & 0.926 \\ 1.320 & 1.086 & 0.926 & 0.0 \end{bmatrix}.$$

The matrix of χ^2 distances between columns is

$$\begin{bmatrix} 0.0 & 0.474 & 0.692 & 0.904 \\ 0.474 & 0.0 & 0.233 & 0.738 \\ 0.692 & 0.233 & 0.0 & 0.660 \\ 0.904 & 0.738 & 0.660 & 0.0 \end{bmatrix}.$$

From the row distances, we see that *Top* and *Above Average* are close in terms of χ^2 distance, while *Above Average* and *Failing* are the furthest apart of the four positions. From the column distances, we see that *Appropriate* and *Fashionable* are close together, while *Appropriate* and *Poorly Dressed* are furthest apart.

We could use multidimensional scaling techniques (Chapter 9) on these distance matrices, or alternatively, cluster analysis techniques (Chapter 8), but instead we concentrate on correspondence analysis.

7.2 Correspondence analysis

Correspondence analysis finds two spaces, one for the rows of $\mathbf{X}$, and one for the columns of $\mathbf{X}$. Points in the row-space represent the rows and points in the column-space represent the columns. The distances between the row-points match, as well as possible, the χ^2 distances between rows. Similarly, the distances between column-points match the χ^2 distances between columns. The two spaces are found using the generalized SVD of $\mathbf{X}$.

Let the generalized SVD of $\mathbf{X}$ be given by

$$\mathbf{X} = \mathbf{A}\mathbf{D}_\lambda\mathbf{B}', \tag{7.2}$$

where

$$\mathbf{A}'\mathbf{D}_R^{-1}\mathbf{A} = \mathbf{I}, \tag{7.3}$$

$$\mathbf{B}'\mathbf{D}_C^{-1}\mathbf{B} = \mathbf{I}. \tag{7.4}$$

The matrix $\mathbf{A}$ is an orthonormal basis for the columns of $\mathbf{X}$, normalized with respect to $\mathbf{D}_R^{-1}$, which allows for the row profile weights $\{R_r\}$. Similarly, $\mathbf{B}$ is an orthonormal basis for the rows of $\mathbf{X}$, normalized with respect to $\mathbf{D}_C^{-1}$, allowing for the column profile weights $\{C_i\}$.

Premultiplying equation (7.2) by $\mathbf{D}_R^{-1}$, the row profiles can be expressed as

$$\mathbf{D}_R^{-1}\mathbf{X} = (\mathbf{D}_R^{-1}\mathbf{A})\mathbf{D}_\lambda \mathbf{B}'. \tag{7.5}$$

Equation (7.3) can be written as

$$(\mathbf{D}_R^{-1}\mathbf{A})'\mathbf{D}_R(\mathbf{D}_R^{-1}\mathbf{A}) = \mathbf{I}. \tag{7.6}$$

Let $\mathbf{U} = \mathbf{D}_R^{-1}\mathbf{A}$, and then equations (7.5) and (7.6) are

$$\mathbf{D}_R^{-1}\mathbf{X} = (\mathbf{U}\mathbf{D}_\lambda)\mathbf{B}',$$
$$\mathbf{U}'\mathbf{D}_R\mathbf{U} = \mathbf{I}.$$

This shows that the rows of $\mathbf{X}$ can be represented as points in the $\mathbf{U}\mathbf{D}_\lambda$ space, with $\mathbf{B}$ the rotation matrix which transforms the points in the $\mathbf{U}\mathbf{D}_\lambda$ space to the row profiles. We can show that the Euclidean distances between points in the $\mathbf{U}\mathbf{D}_\lambda$ space equal the corresponding χ^2 distances between row profiles. To do this, first let $\mathbf{e}_i$ be a vector of zeros, except for the ith element which has the value unity. Then in general, if $\mathbf{Y}$ is an $n \times p$ matrix of coordinates, $\mathbf{e}_i'\mathbf{Y}$ is the row vector of coordinates for the ith point. From equation (7.1) the χ^2 distance d_{rs} between the rth and sth row profiles can be written as

$$d_{rs}^2 = ((\mathbf{e}_r' - \mathbf{e}_s')\mathbf{D}_R^{-1}\mathbf{X}\mathbf{D}_C^{-\frac{1}{2}})((\mathbf{e}_r' - \mathbf{e}_s')\mathbf{D}_R^{-1}\mathbf{X}\mathbf{D}_C^{-\frac{1}{2}})'.$$

Manipulating this equation, we see

$$\begin{aligned} d_{rs}^2 &= (\mathbf{e}_r' - \mathbf{e}_s')\mathbf{D}_R^{-1}\mathbf{X}\mathbf{D}_C^{-1}\mathbf{X}'\mathbf{D}_R^{-1}(\mathbf{e}_r - \mathbf{e}_s) \\ &= (\mathbf{e}_r' - \mathbf{e}_s')(\mathbf{U}\mathbf{D}_\lambda)\mathbf{B}'\mathbf{D}_C^{-1}\mathbf{B}(\mathbf{U}\mathbf{D}_\lambda)'(\mathbf{e}_r - \mathbf{e}_s) \\ &= (\mathbf{e}_r' - \mathbf{e}_s')(\mathbf{U}\mathbf{D}_\lambda)(\mathbf{U}\mathbf{D}_\lambda)'(\mathbf{e}_r - \mathbf{e}_s). \end{aligned}$$

The last term is the squared Euclidean distance between the rth and sth points in the $\mathbf{U}\mathbf{D}_\lambda$ space.

In a similar manner for the column profiles, let $\mathbf{V} = \mathbf{D}_C^{-1}\mathbf{B}$, and

$$\mathbf{D}_C^{-1}\mathbf{X}' = \mathbf{V}\mathbf{D}_\lambda\mathbf{A}',$$
$$\mathbf{V}'\mathbf{D}_C\mathbf{V} = \mathbf{I}.$$

The columns can be represented by points in the $\mathbf{V}\mathbf{D}_\lambda$ space, with $\mathbf{A}$ the rotation matrix to the column profiles. Euclidean distances in the $\mathbf{V}\mathbf{D}_\lambda$ space are equal to the χ^2 distances between the column profiles.

The dimensions of the spaces representing the rows and columns are $\min(n, p)$, and the Euclidean distances between points in the spaces are equal to the χ^2 distances. In order to represent the rows and columns in low-dimensional spaces, only the first few singular values and corresponding singular vectors in the SVD of $\mathbf{X}$ are used.

For the corporate sales data, the various matrices are as follows.

Standardized data matrix:

$$\mathbf{X} = \begin{bmatrix} 0.040 & 0.040 & 0.023 & 0.003 \\ 0.188 & 0.131 & 0.060 & 0.054 \\ 0.067 & 0.141 & 0.107 & 0.087 \\ 0.013 & 0.007 & 0.007 & 0.030 \end{bmatrix}.$$

Matrices, $\mathbf{D}_R$ and $\mathbf{D}_C$:

$$\mathbf{D}_R = \begin{bmatrix} 0.107 & 0.0 & 0.0 & 0.0 \\ 0.0 & 0.433 & 0.0 & 0.0 \\ 0.0 & 0.0 & 0.403 & 0.0 \\ 0.0 & 0.0 & 0.0 & 0.057 \end{bmatrix}$$

$$\mathbf{D}_C = \begin{bmatrix} 0.309 & 0.0 & 0.0 & 0.0 \\ 0.0 & 0.319 & 0.0 & 0.0 \\ 0.0 & 0.0 & 0.198 & 0.0 \\ 0.0 & 0.0 & 0.0 & 0.175 \end{bmatrix}.$$

The row profile matrix:

$$\mathbf{D}_R^{-1}\mathbf{X} = \begin{bmatrix} 0.375 & 0.375 & 0.219 & 0.031 \\ 0.434 & 0.302 & 0.140 & 0.124 \\ 0.167 & 0.350 & 0.267 & 0.217 \\ 0.235 & 0.118 & 0.118 & 0.529 \end{bmatrix}.$$

The column profile matrix:

$$\mathbf{D}_C^{-1}\mathbf{X}' = \begin{bmatrix} 0.130 & 0.126 & 0.119 & 0.019 \\ 0.609 & 0.410 & 0.305 & 0.308 \\ 0.217 & 0.442 & 0.542 & 0.500 \\ 0.043 & 0.021 & 0.034 & 0.173 \end{bmatrix}.$$

The generalized SVD of $\mathbf{X}$ is given by

$$\begin{bmatrix} 0.107 & -0.111 & 0.099 & 0.272 \\ 0.433 & -0.364 & -0.244 & -0.231 \\ 0.403 & 0.350 & 0.317 & -0.132 \\ 0.057 & 0.125 & -0.172 & 0.092 \end{bmatrix} \begin{bmatrix} 1.0 & 0.0 & 0.0 & 0.0 \\ 0.0 & 0.311 & 0.0 & 0.0 \\ 0.0 & 0.0 & 0.218 & 0.0 \\ 0.0 & 0.0 & 0.0 & 0.009 \end{bmatrix}$$

$$\times \begin{bmatrix} 0.309 & -0.360 & -0.259 & 0.129 \\ 0.319 & -0.047 & 0.249 & -0.391 \\ 0.198 & 0.106 & 0.239 & 0.301 \\ 0.174 & 0.300 & -0.229 & -0.039 \end{bmatrix}.$$

The matrices **U** and **V**:

$$U = D_R^{-1}A = \begin{bmatrix} 1.0 & -1.032 & 0.924 & 2.529 \\ 1.0 & -0.842 & -0.563 & -0.533 \\ 1.0 & 0.869 & 0.787 & -0.329 \\ 1.0 & 2.195 & -3.023 & 1.605 \end{bmatrix}$$

$$V = D_C^{-1}B = \begin{bmatrix} 1.0 & 1.165 & -0.840 & 0.419 \\ 1.0 & -0.146 & 0.781 & -1.227 \\ 1.0 & 0.535 & 1.206 & 1.520 \\ 1.0 & 1.722 & -1.310 & -0.224 \end{bmatrix}.$$

The matrices UD_λ and VD_λ:

$$UD_\lambda = \begin{bmatrix} 1.0 & -0.321 & 0.201 & 0.023 \\ 1.0 & -0.262 & -0.123 & -0.005 \\ 1.0 & 0.270 & 0.171 & -0.003 \\ 1.0 & 0.683 & -0.657 & 0.014 \end{bmatrix}$$

$$VD_\lambda = \begin{bmatrix} 1.0 & -0.363 & -0.183 & 0.004 \\ 1.0 & -0.045 & 0.170 & -0.011 \\ 1.0 & 0.167 & 0.262 & 0.014 \\ 1.0 & 0.536 & -0.285 & -0.002 \end{bmatrix}.$$

It can be shown that there is always a singular value of unity, with associated singular vectors **1**. This singular value and the associated singular vectors give rise to the *trivial dimension*. This can be excluded from calculations by removal from the row and column profiles, and using the matrices $D_R^{-1} - 1C'$ and $D_C^{-1} - 1R'$.

As a check that the distances between points in the UD_λ space are equal to the χ^2 distances between rows, the distance between the first point and second point is $\{((-0.321) - (-0.262))^2 + (0.201 - (-0.123))^2 + (0.023 - (-0.005))^2\}^{1/2} = 0.330$, which matches the corresponding χ^2 distance. Similarly for the first and second columns, the distance is $\{((-0.363) - (-0.045))^2 + ((-0.183) - 0.170)^2 + (0.004 - (-0.011))^2\}^{1/2} = 0.475$, which matches the corresponding χ^2 distance (to the second d.p.).

We choose two dimensions in which to display the rows and columns. Figure 7.1 shows the space for the rows, and Figure 7.2 the space for the columns.

There are no surprises in the configuration of points for position, with *Top* and *Above Average* close together, and *Top* and *Failing* furthest apart. For clothes, *Appropriately Dressed* and *Fashionable* are close together. Two spaces, one for rows and one for columns, have been plotted separately, but since the spaces have arisen from the same singular values of **X**, it is possible to plot the rows and column configurations together. The row configuration, UD_λ, can be obtained by transforming the column configuration, VD_λ, and vice versa. To see this, write

$$UD_\lambda = (UD_\lambda)(I) = (UD_\lambda)(B'D_C^{-1}B),$$

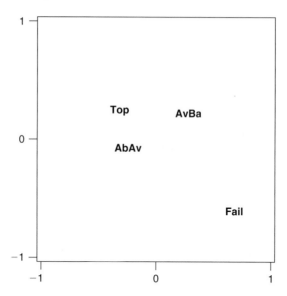

Figure 7.1 Row space

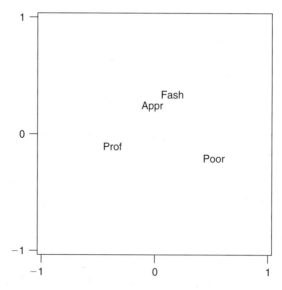

Figure 7.2 Column space

since $\mathbf{B'D}_C^{-1}\mathbf{B} = \mathbf{I}$. Hence

$$
\begin{aligned}
\mathbf{UD}_\lambda &= (\mathbf{UD}_\lambda)\mathbf{B'}(\mathbf{D}_C^{-1}\mathbf{B}) \\
&= (\mathbf{D}_R^{-1}\mathbf{X})\mathbf{V} \\
&= (\mathbf{D}_R^{-1}\mathbf{X})(\mathbf{VD}_\lambda)\mathbf{D}_\lambda^{-1}.
\end{aligned}
$$

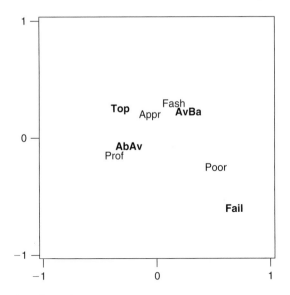

Figure 7.3 Combined configurations

Hence $\mathbf{UD}_\lambda$ can be obtained by premultiplying and postmultiplying $\mathbf{VD}_\lambda$ by $\mathbf{D}_R^{-1}\mathbf{X}$ and $\mathbf{D}_\lambda^{-1}$, respectively. Similarly, $\mathbf{VD}_\lambda$ can be obtained from $\mathbf{UD}_\lambda$ using

$$(\mathbf{VD}_\lambda) = (\mathbf{D}_C^{-1}\mathbf{X}')(\mathbf{UD}_\lambda)\mathbf{D}_\lambda^{-1}.$$

Figure 7.3 shows the two configurations plotted together. Distances between a row point and a column point are not defined, but we note that row points tend to be close to column points for which the row profile values are highest, and vice versa. Thus *Top* and *Average or Below Average* are associated with *Appropriately Dressed* and *Fashionable*. *Above Average* is associated with *Professionally Dressed*, and *Failing* with *Poorly Dressed*.

7.2.1 Inertia

The *total inertia* of the points representing the rows of the matrix $\mathbf{X}$ is a measure of the dispersion of the points. It is the weighted sum of the χ^2 distances of the row points to their centroid. The centroid is given by $\mathbf{R}'\mathbf{D}_R^{-1}\mathbf{X}/\mathbf{R}'\mathbf{1}$. However, $\mathbf{R}'\mathbf{1}=1$ and $\mathbf{R}'\mathbf{D}_R^{-1}=\mathbf{1}'$, and hence the row centroid is given by $\mathbf{1}'\mathbf{X}=\mathbf{C}'$. The total inertia is defined by

$$I = \sum_{r=1}^{n} R_r(\mathbf{R}_r - \mathbf{C})'\mathbf{D}_C^{-1}(\mathbf{R}_r - \mathbf{C}').$$

Now, I can be written as

$$I = \mathrm{tr}(\mathbf{D}_R(\mathbf{D}_R^{-1}\mathbf{X} - \mathbf{1}\mathbf{C}')\mathbf{D}_C^{-1}(\mathbf{D}_R^{-1}\mathbf{X} - \mathbf{1}\mathbf{C}')').$$

Recall, $\mathbf{D}_R^{-1}\mathbf{X} - \mathbf{1}\mathbf{C}'$ is the row profile matrix with the trivial dimension removed. Replace this by $\mathbf{D}_R^{-1}\mathbf{X}$ where it is understood that the trivial dimension has been removed. Then, manipulating the expression for I,

$$I = \mathrm{tr}(\mathbf{D}_R(\mathbf{D}_R^{-1}\mathbf{X})\mathbf{D}_C^{-1}(\mathbf{D}_R^{-1}\mathbf{X})')$$

$$= \mathrm{tr}((\mathbf{A}\mathbf{D}_\lambda\mathbf{B}')\mathbf{D}_C^{-1}(\mathbf{B}\mathbf{D}_\lambda\mathbf{A}')\mathbf{D}_R^{-1})$$

$$= \mathrm{tr}(\mathbf{A}\mathbf{D}_\lambda^2\mathbf{A}'\mathbf{D}_R^{-1})$$

$$= \mathrm{tr}(\mathbf{D}_\lambda^2\mathbf{A}'\mathbf{D}_R^{-1}\mathbf{A})$$

$$= \mathrm{tr}(\mathbf{D}_\lambda^2).$$

Thus the total inertia is equal to the sum of the squared singular values. The dimensions necessary to adequately represent the matrix $\mathbf{X}$ can be judged by the contribution to the total inertia, i.e.

$$\sum_{i=1}^{k} \lambda_i^2 / \sum_{i=1}^{n} \lambda_i^2.$$

A similar derivation can be based on the column profiles.

For the corporate sales data, the inertia values are

λ_i	I	Cumulative
0.3111	0.0968	67%
0.2175	0.0473	100%
0.0089	0.0001	100%

Thus we see that the data can be represented in two dimensions.

7.3 Multiple correspondence analysis

Correspondence analysis is usually used on two-way contingency tables. It can also be used on three-way or higher-way tables by converting these to two-way tables first. For each way (variable), each category for that variable is represented by an indicator variable. So for a particular person (object), zeros are scored for all the categories apart from the one observed, which itself scores one. For example, if a corporate sales woman for the data above was *Appropriately Dressed*, then her scores for the four indicator variables for *Dress* are: *Prof* – 0, *Appr* – 1, *Fash* – 0, and *Poor* – 0. Each variable is converted into a sequence of indicator variables and then all are placed in a grand indicator matrix, $\mathbf{Z}$. Thus for the corporate sales data, the columns of the matrix are:

P1   P2   P3   P4   Prof   Appr   Fash   Poor.

The matrix **Z** is

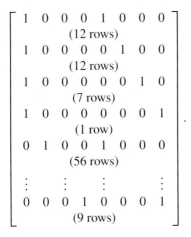

$$
\begin{bmatrix}
1 & 0 & 0 & 0 & 1 & 0 & 0 & 0 \\
& & & \text{(12 rows)} & & & & \\
1 & 0 & 0 & 0 & 0 & 1 & 0 & 0 \\
& & & \text{(12 rows)} & & & & \\
1 & 0 & 0 & 0 & 0 & 0 & 1 & 0 \\
& & & \text{(7 rows)} & & & & \\
1 & 0 & 0 & 0 & 0 & 0 & 0 & 1 \\
& & & \text{(1 row)} & & & & \\
0 & 1 & 0 & 0 & 1 & 0 & 0 & 0 \\
& & & \text{(56 rows)} & & & & \\
\vdots & & \vdots & & \vdots & & & \vdots \\
0 & 0 & 0 & 1 & 0 & 0 & 0 & 1 \\
& & & \text{(9 rows)} & & & &
\end{bmatrix}.
$$

Example

The four-way contingency table of tumours in mice in Table 14.7 of Chapter 14 was converted into a 401×8 indicator matrix, **Z**. The columns were

Strain	Sex	Exposure	Tumour
X Y	M F	E C	T NT

Correspondence analysis was carried out on **Z**, resulting in non-zero singular values of 1.0 (trivial dimension), 0.552, 0.502, 0.489, and 0.452. Figure 7.4 shows the eight column variables plotted using the first two dimensions. We see that control, C, and tumour, T, are close together, and that exposed, E, and no tumour NT stand together away from the other points. This is in accordance with the findings of Chapter 14, that exposure to Avadex appears to decrease the risk of a tumour occurring.

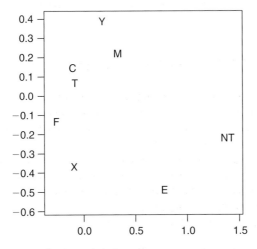

Figure 7.4 Multiple correspondence analysis for mice

Note that if the data for a two-way contingency table are converted to a grand indicator matrix, **Z**, and then correspondence analysis carried out, the results will not be the same as if the analysis had been done on the original two-way contingency table, **X**. However, the two analyses are closely related; see Greenacre (1993) for further details.

7.4 Exercises

1. The following table gives the results of a study of malignant melanoma, carried out by Roberts et al. (1981), where site of tumour (*head/neck* (h), *trunk* (t), *extremities* (e)) and histological type (*Hutchinson's melanotic freckle* (H), *superficial spreading melanoma* (S), *Nodular* (N), and *Indeterminate* (I)) were recorded for four hundred patients.

Type	Site		
	h	t	e
H	22	2	10
S	16	54	115
N	19	33	73
I	11	17	28

Carry out a correspondence analysis on these data.

2. Carry out a correspondence analysis using the grand indicator matrix, **Z**, in Section 7.3 for the corporate sales data.

3. The following table gives the number of survivors and non-survivors for the persons sailing on the *Titanic* on its tragic inaugural journey. The persons are classified by *Class, Age, Sex,* and *Survival*. The data are available at the website http://www.statsci.org/data/general/titanic.html, and also others. It is also a JMP sample dataset.

Class	Age	Sex	Survived	Died
crew	adult	F	20	3
crew	adult	M	192	670
1st	adult	F	140	4
1st	adult	M	57	118
1st	child	F	1	0
1st	child	M	5	0
2nd	adult	F	80	13
2nd	adult	M	14	154
2nd	child	F	13	0
2nd	child	M	11	0
3rd	adult	F	76	89
3rd	adult	M	75	387
3rd	child	F	14	17
3rd	child	M	13	35

Calculate the indicator matrix **Z** from these data, using the binary variables (crew, first, second, third, adult, child, female, male, survived, died). Carry out a correspondence analysis using **Z**.

8

Cluster analysis

Suppose a company manufacturing personal care products has data on a sample of its consumers, and wishes to use these data in order to place them into groups. The data might include variables relating to lifestyle, habits and income. Consumers placed within a group would be similar in some way. As another example, suppose data are collected on 100 British universities; data such as degree programmes offered, degree results, amount spent per student, library provision, careers entered, etc. It may be a useful exercise to try to group the universities so that the universities within a group are similar. We say the universities are to be *classified* into *groups* or *classes*. Other terms are to say the universities are to be *clustered*, or the consumers are to be *segmented*. Classifying biological organisms has had a long history and is generally known as *taxonomy*. Note, in cluster analysis, there are no classes *a priori*, where we are attempting to place each object into one of them, as with discriminant analysis. The data themselves give rise to the classes or groups, hopefully in a meaningful manner.

Suppose a cluster analysis is to be carried out on N objects. Data are collected on each object which can be in the form of a random vector or dissimilarity data. There are several methods of cluster analysis. These include, *hierarchical methods, optimizing methods, mixture models,* and *density methods*. Not all techniques can be covered here, and the reader is referred to other texts such as Everitt et al. (2001) or Gordon (1999) for a fuller account. Note that with some methods it is also possible to cluster the variables as opposed to the objects, with the role of objects and variables interchanged.

8.1 Hierarchical methods

Hierarchical methods of cluster analysis start with each object in a cluster of its own (i.e. N clusters) and then continually join clusters together, until there is only one cluster consisting of all the objects. Clusters are joined on the basis of 'shortest distance' between clusters. Alternatively, we can start with one cluster of all the objects and then split this cluster into more and more clusters. The first method is the most common of the two.

A set of clusters is arrived at which is usually illustrated with a *dendrogram* as illustrated in Figure 8.1 below. The dendrogram continually branches from the top, with the final branches at the bottom leading to the objects that are being clustered. Viewing the dendrogram from the bottom and working upwards, the dendrogram starts with seven singletons {1}, {2}, {3}, {4}, {5}, {6}, and {7}. Then moving upwards, clusters are seen to be amalgamated at the nodes of the dendrogram, with singleton clusters being joined together, or a singleton joining

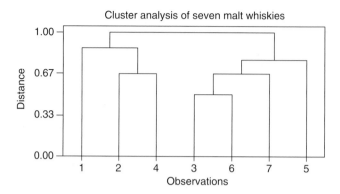

Figure 8.1 Dendrogram of seven malt whiskies using complete linkage: 1 – Glenburgie, 2 – Strathisla, 3 – Balblair, 4 – Clynelish, 5 – Royal Brakla, 6 – Teaninich, 7 – Glen Garioch

a larger cluster, or larger clusters being amalgamated. The scale to the left of the dendrogram gives the distances between clusters or single objects at the point at which they are joined. Choosing a particular value d of the distance between clusters above which amalgamation of clusters is deemed no longer to occur, will define a set of clusters. From Figure 8.1 with $d = 0.8$, the individual clusters are formed by those objects that emanate from each of the branches cut by a horizontal line drawn at d. The different ways of measuring distance between clusters gives rise to different dendrograms. The following are some of the more popular ways of defining such distances.

8.1.1 Complete linkage or furthest neighbour method

This is a well used method for cluster analysis and will serve the purpose of illustrating the method in detail. Cox and Cox (2000) analyse some binary data collected on malt whiskies. Each whisky is described as to its nose and taste characteristics, such as 'delicate', 'peaty', 'dry' and 'smokey'. There are twenty-six characteristics in total, each represented by a binary variable. There are nineteen whiskies, each scoring 1 for a characteristic if it is present in the whisky, and 0 otherwise. The data can be found in the book by Cox and Cox.

The first seven whiskies, labelled $1, 2, \ldots, 7$, have been chosen to illustrate the complete linkage method. We start with seven singleton clusters each containing one of the whiskies. For each pair of whiskies the Jaccard coefficient, s_{rs}, was calculated, and then transformed into a dissimilarity, $d_{rs} = 1 - s_{rs}$. The dissimilarities are used to define the distance between clusters. Below is the matrix consisting of these distances.

$$
\begin{bmatrix}
 & 1 & 2 & 3 & 4 & 5 & 6 & 7 \\
1 & 0.0 \\
2 & 0.875 & 0.0 \\
3 & 1.000 & 0.667 & 0.0 \\
4 & 0.875 & 0.667 & 1.000 & 0.0 \\
5 & 0.917 & 0.800 & 0.500 & 0.909 & 0.0 \\
6 & 0.889 & 0.714 & 0.500 & 0.875 & 0.700 & 0.0 \\
7 & 1.000 & 1.000 & 0.600 & 1.000 & 0.778 & 0.667 & 0.0
\end{bmatrix}
$$

The two clusters with shortest distance between them are now joined into one cluster. Here the shortest distance is 0.500 between clusters (3) and (5), but also between clusters (3) and (6). We arbitrarily choose to join clusters (3) and (6). Denote this cluster of size two by (3, 6). New distances, $d_{r(3,6)}$, between each of the singleton clusters and this cluster are found as

$$d_{r(3,6)} = \max(d_{r3}, d_{r6}) \quad r = 1, 2, 4, 5, 7. \tag{8.1}$$

The other distances remain the same. The new distance matrix is

$$
\begin{bmatrix}
 & 1 & 2 & (3,6) & 4 & 5 & 7 \\
1 & 0.0 \\
2 & 0.875 & 0.0 \\
(3,6) & 1.000 & 0.714 & 0.0 \\
4 & 0.875 & 0.667 & 1.000 & 0.0 \\
5 & 0.917 & 0.800 & 0.700 & 0.909 & 0.0 \\
7 & 1.000 & 1.000 & 0.667 & 1.000 & 0.778 & 0.0
\end{bmatrix}.
$$

The smallest distance is now 0.667 between clusters (2) and (4), and between (3, 6) and (7). We join (2) and (4) and calculate new distances as before. The new distance matrix is

$$
\begin{bmatrix}
 & 1 & (2,4) & (3,6) & 5 & 7 \\
1 & 0.0 \\
(2,4) & 0.875 & 0.0 \\
(3,6) & 1.000 & 1.000 & 0.0 \\
5 & 0.917 & 0.909 & 0.700 & 0.0 \\
7 & 1.000 & 1.000 & 0.667 & 0.778 & 0.0
\end{bmatrix}.
$$

Next cluster (7) is added to the cluster (3, 6) at a distance of 0.667 to form the cluster (3, 6, 7). The new distance matrix is

$$
\begin{bmatrix}
 & 1 & (2,4) & (3,6,7) & 5 \\
1 & 0.0 \\
(2,4) & 0.875 & 0.0 \\
(3,6,7) & 1.000 & 1.000 & 0.0 \\
5 & 0.917 & 0.909 & 0.778 & 0.0
\end{bmatrix}.
$$

Next cluster (5) is added to the cluster (3, 6, 7) at a distance of 0.778. The new distance matrix is

$$
\begin{bmatrix}
 & 1 & (2,4) & (3,5,6,7) \\
1 & 0.0 \\
(2,4) & 0.875 & 0.0 \\
(3,5,6,7) & 1.000 & 1.000 & 0.0
\end{bmatrix}.
$$

Cluster (1) is now added to the cluster (2, 4) at a distance of 0.875. The updated distance matrix is

$$
\begin{bmatrix}
 & (1,2,4) & (3,5,6,7) \\
(1,2,4) & 0.0 \\
(3,5,6,7) & 1.000 & 0.0
\end{bmatrix}.
$$

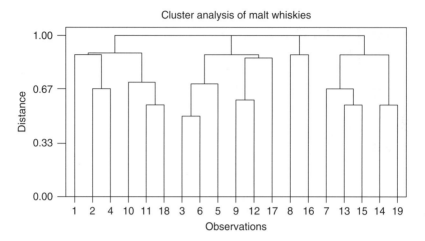

Figure 8.2 Dendrogram of nineteen whiskies

Finally these last two clusters are joined at a distance of 1.0.

Figure 8.1 shows the dendrogram for these whiskies. The scale on the left shows at which distances the various clusters were formed. If a distance of 0.9 were chosen to split the seven whiskies into two clusters, then these clusters are $(1, 2, 4)$ and $(3, 5, 6, 7)$. Out of interest, whiskies 1 and 2 are from Speyside, whiskies $3, 4, 5, 6$ are from the Northern Highlands and 7 is from the Eastern Highlands.

Figure 8.2 shows the dendrogram for all nineteen whiskies, again using the complete linkage method. The whiskies are: *1 – Glenburgie, 2 – Strathisla, 3 – Balblair, 4 – Clynelish, 5 – Royal Brakla, 6 – Teaninich, 7 – Glen Garioch, 8 – Glenturret, 9 – Oban, 10 – Bladnoch, 11 – Littlemill, 12 – Ardbeg, 13 – Bowmore, 14 – Lagavulin, 15 – Laphroaig, 16 – Highland Park, 17 – Isle of Jura, 18 – Tobermory, 19 – Bushmills.* If four clusters are to be chosen, then these would be $\{1, 2, 4, 10, 11, 18\}$, $\{3, 6, 5, 9, 12, 17\}$, $\{8, 16\}$, and $\{7, 13, 15, 14, 19\}$. The first cluster has two whiskies from Speyside, two from the Lowlands, one from the Highlands and one from the Islands; the second cluster has four whiskies from the Highlands, one from Islay and one from the Islands, the third cluster has one whisky from the Highlands and the other from the Islands; the fourth cluster has three whiskies from Islay, one from the Highlands and one from Northern Ireland.

8.1.2 Single linkage or nearest neighbour method

For the complete linkage method, the distance between two clusters was the distance between the furthest apart pair of members, one from each cluster. The single linkage method defines the distance between two clusters as the distance between the nearest pair of members, one from each cluster. The procedure for the single linkage method is the same as that for the complete linkage method, except (8.1) is replaced by

$$d_{r(3,6)} = \min(d_{r3}, d_{r6}) \quad r = 1, 2, 4, 5, 7.$$

Figure 8.3 shows the dendrogram for the whisky data, using the single linkage method. Note the difference to that for the complete linkage method. In general, the single linkage

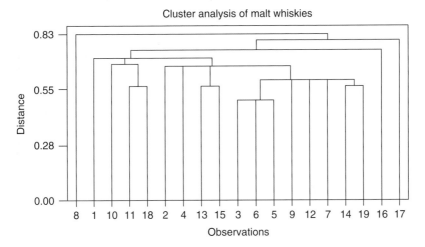

Figure 8.3 Dendrogram using single linkage

method will join two clusters which might be close, simply because a member of one cluster is close to a member of another cluster, but all other pairs of members, one from each cluster, might be far apart. The single linkage method can give rise to 'chaining', where one singleton cluster joins a larger cluster, then another singleton cluster joins the same large cluster, then another, and so on. The complete linkage method guards against this, but it can lead to clusters of similar sizes being formed, when perhaps this is not appropriate.

8.1.3 Other linkages

We now denote specific clusters by R, S and T. For the complete linkage method, the distance between cluster T and the cluster formed by joining clusters R and S, is defined as

$$d_{T(R,S)} = \max(d_{TR}, d_{TS}).$$

For the single linkage method the distance is

$$d_{T(R,S)} = \min(d_{TR}, d_{TS}).$$

Many other methods of defining distances between clusters have been proposed. The *centroid linkage* measures distance between clusters as the distance between their centroids. At each step, the two clusters with the smallest distance between centroids are amalgamated. *Ward's method* merges the two clusters which give rise to the smallest increase in the sum of squared within-cluster distances. Lance and Williams (1967) proposed a flexible measure for the distance $d_{T(R,S)}$ defined as

$$d_{T(R,S)} = \alpha_R d_{TR} + \alpha_S d_{TS} + \beta d_{RS} + \gamma |d_{TR} - d_{TS}|,$$

where α_R, α_S, β and γ are parameters to be chosen. For instance, if $\alpha_R = \alpha_S$, $\beta = 0$, $\gamma = -\frac{1}{2}$, then

$$d_{T(R,S)} = \tfrac{1}{2}d_{TR} + \tfrac{1}{2}d_{TS} - \tfrac{1}{2}|d_{TR} - d_{TS}|.$$

If $d_{TR} > d_{TS}$, then $d_{T(R,S)} = d_{TS}$, and if $d_{TR} < d_{TS}$, then $d_{T(R,S)} = d_{TR}$. Thus $d_{T(R,S)} = \max(d_{TR}, d_{TS})$, which is the definition used for the complete linkage method. Similarly, for $\alpha_R = \alpha_S$, $\beta = 0$, $\gamma = +\frac{1}{2}$, then $d_{T(R,S)} = \min(d_{TR}, d_{TS})$, giving rise to the single linkage method. The required parameter values for other linkages are given in Everitt et al. (2001), Gordon (1999) and Krzanowski and Marriott (1995), as well as the original paper by Lance and Williams (1967).

8.2 *K*-means clustering

K-means clustering covers a group of techniques that find clusters by optimizing various criteria. Suppose we have n observations for p-dimensional data, which have been grouped into K clusters by some means. Denote the observations by $\mathbf{x}_{Ri}$, $(R = 1, \ldots, K; i = 1, \ldots, n_R)$, where R refers to cluster R, and i refers to the ith observation in cluster R. Considering the clusters as residing in p-dimensional space, the location of cluster R in this space can be measured by its cluster mean value, $\bar{\mathbf{x}}_R$,

$$\bar{\mathbf{x}}_R = \frac{1}{n_R} \sum_{i=1}^{n_R} \mathbf{x}_{Ri}.$$

The location of the entire dataset is measured by the overall mean, $\bar{\mathbf{x}}$,

$$\bar{\mathbf{x}} = \frac{1}{n} \sum_{R=1}^{K} \sum_{i=1}^{n_R} \mathbf{x}_{Ri}.$$

The overall dispersion of the data is measured by the matrix

$$\mathbf{T} = \sum_{R=1}^{K} \sum_{i=1}^{n_R} (\mathbf{x}_{Ri} - \bar{\mathbf{x}})(\mathbf{x}_{Ri} - \bar{\mathbf{x}})'.$$

This can be split into the dispersion within the clusters, $\mathbf{W}$, and the dispersion between the clusters, $\mathbf{B}$, as in MANOVA (see Chapter 11). So

$$\mathbf{T} = \mathbf{W} + \mathbf{B},$$

where

$$\mathbf{W} = \sum_{R=1}^{K} \sum_{i=1}^{n_R} (\mathbf{x}_{Ri} - \bar{\mathbf{x}}_R)(\mathbf{x}_{Ri} - \bar{\mathbf{x}}_R)'$$

$$\mathbf{B} = \sum_{R=1}^{K} n_R (\bar{\mathbf{x}}_R - \bar{\mathbf{x}})(\bar{\mathbf{x}}_R - \bar{\mathbf{x}})'. \tag{8.2}$$

Ideally we would like to have the clusters such that $\mathbf{W} = \mathbf{0}$, with all the dispersion in the data accounted for by the clusters. However, in practice, we can only hope that the elements of $\mathbf{W}$ are small compared to those of $\mathbf{B}$. K-means clustering attempts to group the observations so that some criterion measuring the size of $\mathbf{W}$ is optimized. Various criteria used are tr $\mathbf{W}$

to be minimized, $|\mathbf{W}|$ to be minimized, and $\mathrm{tr}(\mathbf{BW}^{-1})$ to be maximized. For large datasets, the number of possible partitions of the data into K groups becomes an unwieldy number, and so a complete enumeration is not feasible. Instead, a minimization (maximization) algorithm has to be used. The steps are:

1. Find an initial partition of the data into K-groups, either randomly assigning the observations, or using another clustering method, for example complete linkage.
2. Calculate the score for the value of the criterion being used, for example $|\mathbf{W}|$.
3. For each observation, $\mathbf{x}_{Ri}$, calculate whether the score would be reduced (increased) if $\mathbf{x}_{Ri}$ were transferred to another group. If so, transfer the observation that reduces (increases) the score the most.
4. Return to 2 unless the score cannot be reduced (increased) any further.

This procedure will find a local minimum (maximum) for the score, but it is not necessarily the global minimum (maximum). It is best to repeat the procedure several times, using different starting partitions, and hoping the global minimum (maximum) will be found, or at least a local minimum (maximum) close to the global value.

K-means clustering needs the number of clusters to be stated. Knowing the number of clusters to choose could be difficult. The dendrogram from a hierarchical method used on the data might indicate the number to choose. Choosing a range of number of clusters and using K-means clustering each time, may show a good value to be used. Along these lines, various objective criteria have been suggested. For example Marriott (1971) suggests choosing K as the value which minimizes $K^2|\mathbf{W}|$, while Hartingan (1975) suggests using

$$c_K = (\mathrm{tr}\,\mathbf{W}_K/\mathrm{tr}\mathbf{W}_{K+1} - 1)(\text{no. rows of } \mathbf{X} - K - 1),$$

where $\mathbf{W}_K$ is the dispersion within clusters matrix when there are K groups. As a guide, if $c_K > 10$ then the extra cluster is added.

Example

The second exercise of Chapter 9 gives percentages for the grades obtained in twenty-four subjects in GCSE examinations. These were subjected to K-means clustering, having first scaled the data so that the sum over the subjects, for each grade, equalled one hundred. This was to put each grade A to G on an equal footing, otherwise the grades with the largest variation would dominate the clustering. Four clusters were chosen based on Hartingan's criterion. S-Plus was used for the analysis. The clusters were:

Cluster 1: *Business Studies, Computer Studies, CDT, Home Economics Mathematics, Science, Social Sciences*
Cluster 2: *Classical Civilization, German, Spanish, Music*
Cluster 3: *Art and Design, Economics, English, English Literature, Geography, History, French, Religious Studies, Biology, Physics*
Cluster 4: *Latin, Greek*

The cluster centres are located at

Cluster 1	3.2	4.4	4.7	4.7	4.3	3.9	3.4
Cluster 2	2.3	3.2	4.2	4.6	5.2	5.5	5.8
Cluster 3	5.6	5.2	3.9	3.5	3.1	3.5	3.9
Cluster 4	14.0	5.5	2.3	1.2	0.7	0.7	0.7.

8.3 Density search methods

If we think of our observations as n points sitting in a p-dimensional metric space, then clusters could be considered as those points residing in areas of high density. This would usually leave some of the points not within a cluster, i.e. those points in sparse areas of the space. Various measures of density can be formulated. For instance, from an arbitrary point P, with distances to the k nearest observations, $d_1, \ldots, d_k$, a measure of density at P is $(\bar{d})^{-2}$, where $\bar{d}$ is the mean of $d_1, \ldots, d_k$. Clusters are formed by those observations within areas of high density, defined by some threshold value. A grid of points covering the space containing the data points could be used to find these areas of low and high density. For high-dimensional data this may be prohibitive, and so density might only be measured at the actual data points in the space. Another approach is to use a hypersphere of radius R, placed at each observation, and to count the number of neighbouring observations that lie within the hypersphere. Observations are labelled dense if they have k or more neighbouring observations

8.4 Self-organizing maps

In the context of machine learning, Kohonen (1982, 1990) introduced the self-organizing map (SOM). SOMs have strong links to K-means clustering. In their simplest form, SOMs map multidimensional data onto nodes formed by vertices of a low-dimensional grid (usually two-dimensional). These nodes and their associated observations can be considered as clusters. Figure 8.4 shows such a map consisting of five nodes. Let the nodes be denoted by $n_j, (j = 1, \ldots, k)$. The node, n_j, has a vector associated with it, $\mathbf{m}_j$. Also shown on the map are the associated observations for each node.

Initially all vertices of the grid start as nodes and are given node vectors $\mathbf{m}_i$ chosen randomly as one of observation vectors, $\mathbf{x}_i$. The map is updated as each observation is mapped onto one of the nodes, whereupon the node vectors are updated. The observations are mapped sequentially with updating occurring at each step. For each observation, $\mathbf{x}_i$, the distance, $d(\mathbf{x}_i, \mathbf{m}_j)$, between it and each node vector is calculated. Distance can be measured in several ways, for example, by Euclidean distance, $d(\mathbf{x}_i, \mathbf{m}_j) = \{(\mathbf{x}_i - \mathbf{m}_j)'(\mathbf{x}_i - \mathbf{m}_j)\}^{\frac{1}{2}}$. The observation is then mapped onto the node which has minimum distance, say node c. The node vector $\mathbf{m}_c$, together with the node vectors for neighbouring nodes, are then updated. Let N_c denote the set of neighbouring nodes of node c (including node c itself), which might be defined as those nodes within a distance d_β of node c. The node vector $\mathbf{m}_c$ for node c and those in its neighbourhood are then updated to

$$\mathbf{m}_j + \alpha h_{cj}(\mathbf{x}_i - \mathbf{m}_j), \ (j \in N_c)$$

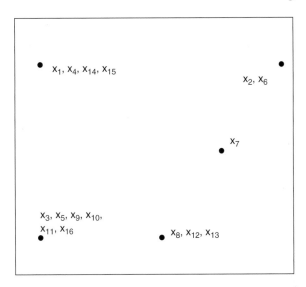

Figure 8.4 SOM with five nodes

where $0 < \alpha < 1$ and h_{cj} is a weight which gives more emphasis to those nodes close to node c.

All observations are sequentially applied to the map, whereupon, when the last one has been allocated and the node vectors updated, the process is repeated, and then repeated again and again. However, as the process evolves, the neighbourhood of each node and the value of α decrease, which will eventually lead to convergence of the map. One method of achieving this is to choose the node vector updating equation as

$$\mathbf{m}_j^{(t+1)} = \mathbf{m}_j^{(t)} + \alpha(t) \exp\left\{(-d_{cj}^2)/\beta(t)\right\} (\mathbf{x}_i - \mathbf{m}_j),$$

where d_{cj} is the Euclidean distance between nodes c and j in the map, and $\alpha(t)$ and $\beta(t)$ are monotonically decreasing functions of t, the tth iteration of the map. For instance $\alpha(t) = \exp(-t/1000)$, $\beta(t) = \exp(-t/1000)$ might be chosen.

Example

A self-organizing map was formed for the GCSE data for 1991. Figure 8.5 shows the map, where the algorithm has produced sixteen nodes and then clustered these into four groups:

Cluster 1: *Business Studies, Mathematics, Science, Home Economics, CDT, Social Sciences, French, Computer Studies*
Cluster 2: *English, English Literature, Physics, Economics, Chemistry, Music, Biology, Art and Design, Geography, History, German, Religious Studies, Spanish*
Cluster 3: *Classical Civilization*
Cluster 4: *Latin, Greek*

Compare these groupings to those obtained from K-means clustering.

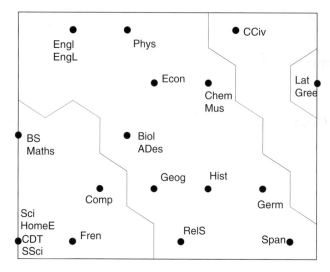

Figure 8.5 SOM of the 1991 GCSE data.

8.5 Model based methods

8.5.1 A likelihood method

Suppose data consist of n observations on random vector $\mathbf{x}$ with each observation arising from one of g populations. Each population represents a cluster with associated pdf for $\mathbf{x}$ denoted by $f_i(\mathbf{x})$ $(i = 1, \ldots, g)$. However, we do not know from which population each observation has arisen.

Let G_i be the set of observations from the ith cluster (even though we do not know which these are). Then the likelihood is

$$L = \prod_{i_1 \in G_1} f_1(\mathbf{x}_{i_1}) \prod_{i_2 \in G_2} f_2(\mathbf{x}_{i_2}) \ldots \prod_{i_g \in G_g} f_g(\mathbf{x}_{i_g}).$$

The likelihood is maximized over the parameters of the density functions and also over cluster membership. This can be a very intensive computing task.

Example

For the ith cluster, let $\mathbf{x} \sim N_p(\boldsymbol{\mu}_j, \boldsymbol{\Sigma}_j)$. The likelihood is

$$\prod_{i=1}^{g} \prod_{j=1}^{n_i} \frac{1}{(2\pi)^{p/2}|\boldsymbol{\Sigma}_i|^{1/2}} \exp\left\{-\tfrac{1}{2}(\mathbf{x}_{ij} - \boldsymbol{\mu}_i)'\boldsymbol{\Sigma}_i^{-1}(\mathbf{x}_{ij} - \boldsymbol{\mu}_i)\right\}.$$

The log-likelihood is

$$\text{constant} - \frac{1}{2} \sum_{i=1}^{g} \sum_{j=1}^{n_i} (\mathbf{x}_{ij} - \boldsymbol{\mu}_i)' \boldsymbol{\Sigma}_i^{-1} (\mathbf{x}_{ij} - \boldsymbol{\mu}_i) - \frac{1}{2} \sum_{i=1}^{g} n_i \log |\boldsymbol{\Sigma}_i|. \qquad (8.3)$$

For given cluster membership, the maximum likelihood estimates of $\boldsymbol{\mu}_i$ and $\boldsymbol{\Sigma}_i$ are given by $\hat{\boldsymbol{\mu}} = \bar{\mathbf{x}}_i$ (see Chapter 3), $\hat{\boldsymbol{\Sigma}}_i = n_i^{-1} \sum_{j=1}^{n_i} (\mathbf{x}_i - \bar{\mathbf{x}})(\mathbf{x}_j - \bar{\mathbf{x}})^T$. The value of the log-likelihood at its maximum is then

$$\text{constant} - \frac{1}{2} \sum_{i=1}^{g} n_i \log |\hat{\boldsymbol{\Sigma}}_i|. \qquad (8.4)$$

However, we do not know the cluster membership, and so (8.3) has to be maximized over all possible partitions of the observations into g clusters, where g can vary from 1 to n. Note, however, that the estimate of $\boldsymbol{\Sigma}_i$ and the log-likelihood will become degenerate if the number of observations in the ith cluster is less than $p + 1$.

In practice it will usually be impossible to calculate the log-likelihood for all possible partitions of the observations into clusters, ranging in number from 1 to n, and so an iterative approach has to be used. The log-likelihood can be found for a single cluster of size n. Then the observations can be randomly split into two clusters, and the log-likelihood found. Then, in turn, each observation is transferred from its cluster into the other, and the log-likelihood found again. If the log-likelihood has increased, then the change of cluster is retained, and otherwise not. The changing cluster for each observation is repeated several times until the likelihood no longer increases. The whole procedure is repeated for three clusters, then four, and so on. A global maximum of the log-likelihood is not guaranteed to be found, and it is recommended that several random starting cluster memberships be used.

8.5.2 A mixture model

Another modelling approach to clustering is to assume that the observations come from a mixture distribution, each component in the mixture representing one of the clusters. Here, we briefly consider a mixture of multivariate normal distributions, with probability density function,

$$f(\mathbf{x}) = \sum_{i=1}^{g} \lambda_i f_i(\mathbf{x}_i; \boldsymbol{\mu}_i, \boldsymbol{\Sigma}_i),$$

where $f_i(\mathbf{x}_i; \boldsymbol{\mu}_i, \boldsymbol{\Sigma}_i)$ is the MVN probability density function for the ith cluster, and λ_i is its mixture weight, with $\sum_{i=1}^{g} \lambda_i = 1$.
The log-likelihood is given by

$$l = \sum_{j=1}^{n} \log \left\{ \sum_{i=1}^{g} \lambda_i f_i(\mathbf{x}_j; \boldsymbol{\mu}_i, \boldsymbol{\Sigma}) \right\}.$$

The maximum likelihood estimates of λ_i, $\boldsymbol{\mu}_i$, and $\boldsymbol{\Sigma}$ can be found using the EM algorithm (Expectation-Minimization algorithm, see Dempster et al., 1977), and then cluster membership is determined for each observation according to the largest value of

$$\frac{\hat{\lambda} f_i(\hat{\mu}_i, \hat{\Sigma}_i)}{\sum_{j=1}^{g} \hat{\lambda} f_j(\hat{\mu}_j, \hat{\Sigma}_j)} .$$

8.6 Exercises

1. For the dissimilarity matrix produced for the pets data in Exercise 1.2, draw a dendrogram by hand using (i) the nearest neighbour method, and then (ii) the furthest neighbour method.

2. Use a statistical software package to generate six independent columns of twenty observations, each from a uniform distribution on the interval [0, 1]. Use the package to carry out various clustering techniques, both on the observations, 1–20, and the variables, 1–6.
   Repeat the exercise, generating a new set of data. Compare the clusters obtained from the two occasions. What do you conclude?

3. From the likelihood of (8.3), show that it has maximum value as given in (8.4).

4. Carry out a cluster analysis on the Chania data described in Chapter 1, on both the observations and on the variables.

9

Multidimensional scaling

Multidimensional scaling (MDS) covers a variety of techniques, aimed at representing objects (or individuals) by a configuration of points in a space, usually Euclidean, where each point represents one of the objects. Objects which are 'similar', in some sense, are expected to have their respective points in the space close to each other, while objects which are dissimilar would have their points further apart. Measurement of similarity or dissimilarity between objects is made using one of the similarity/dissimilarity measures discussed in Chapter 1. Multidimensional scaling covers techniques such as classical scaling, nonmetric scaling, Procrustes analysis, biplots, unfolding, correspondence analysis and individual differences scaling. Of these, biplots and correspondence analysis have been covered in Chapters 6 and 7, while this chapter covers classical scaling, nonmetric scaling, Procrustes analysis and individual differences scaling. Books that give detailed coverage of MDS include Borg and Groenen (1997) and Cox and Cox (2000).

9.1 Classical scaling

First, we pose the following problem, which appears in the first instance not to have anything to do with MDS, nor statistics in general, but underlies the concept of multidimensional scaling. Given the distances between all possible pairs of twelve French cities, but not their coordinates, can we reconstruct a map of the cities? Of course it does not have to be French cities, and it could be any other number of them apart from twelve. The answer is yes, and to illustrate the method of reconstruction we will use twelve such cities, with the distances between them shown in Table 9.1.

9.1.1 Reconstruction of the map

The following algorithm is used to construct a map of the cities. Consider n points in a p-dimensional space, with coordinates $\mathbf{x}_r$ $(r = 1, \ldots, n)$, where $\mathbf{x}_r = (x_{r1}, \ldots, x_{rp})'$. The squared Euclidean distance between points r and s is

$$d_{rs}^2 = (\mathbf{x}_r - \mathbf{x}_s)'(\mathbf{x}_r - \mathbf{x}_s). \tag{9.1}$$

The data we work with are the set of squared distances, $\{d_{rs}^2\}$. Our task is to find the set of coordinates, $\{\mathbf{x}_r\}$, from these. The algorithm is in two parts. First the inner product matrix,

Table 9.1 Distances (mm) between twelve French cities: scale 1 mm : 2.6 km

	Bord	Boul	Bour	Caen	Dijo	Lyon	Mars	Nants	Pari	Renn	Roue	Toul
Bordeaux	0	.	.	.	.	.	.	.	.	.	.	.
Boulogne	169	0	.	.	.	.	.	.	.	.	.	.
Bourges	84	100	0	.	.	.	.	.	.	.	.	.
Caen	120	55	78	0	.	.	.	.	.	.	.	.
Dijon	128	111	50	112	0	.	.	.	.	.	.	.
Lyon	118	150	60	136	44	0	.	.	.	.	.	.
Marseille	125	218	118	195	110	67	0	.	.	.	.	.
Nantes	68	113	74	59	123	128	171	0	.	.	.	.
Paris	124	52	49	50	65	99	163	85	0	.	.	.
Rennes	93	92	81	38	126	140	189	25	77	0	.	.
Rouen	131	36	69	27	93	124	187	78	27	63	0	.
Toulouse	52	200	99	157	124	89	78	115	146	138	161	0

B, is found, where

$$\mathbf{B} = [\mathbf{x}_r'\mathbf{x}_s],$$

and secondly, from **B**, we extract the coordinates $\mathbf{x}_r$.

To find B.   There can be no unique solution to our problem of finding the coordinates of the points from the pairwise distances between them. This is because any solution giving the coordinates can be arbitrarily translated, rotated and reflected, to another possible solution – the distances between the pairs of points will stay the same! Without loss of generality, we find a solution that has its centroid at the origin, so that

$$\sum_{r=1}^{n}\mathbf{x}_r = \mathbf{0}.$$

From equation (9.1)

$$d_{rs}^2 = \mathbf{x}_r'\mathbf{x}_r + \mathbf{x}_s'\mathbf{x}_s - 2\mathbf{x}_r'\mathbf{x}_s. \tag{9.2}$$

Summing equation (9.2) over r and dividing by n gives

$$\frac{1}{n}\sum_{r=1}^{p}d_{rs}^2 = \frac{1}{n}\sum_{r=1}^{n}\mathbf{x}_r'\mathbf{x}_r + \frac{1}{n}\sum_{r=1}^{n}\mathbf{x}_s'\mathbf{x}_s - \frac{2}{n}\left(\sum_{r=1}^{n}\mathbf{x}_r'\right)\mathbf{x}_s$$

$$= \frac{1}{n}\sum_{r=1}^{n}\mathbf{x}_r'\mathbf{x}_r + \mathbf{x}_s'\mathbf{x}_s.$$

Similarly, summing equation (9.2) over s and dividing by n, gives

$$\frac{1}{n}\sum_{s=1}^{p}d_{rs}^2 = \mathbf{x}_r'\mathbf{x}_r + \frac{1}{n}\sum_{s=1}^{n}\mathbf{x}_s'\mathbf{x}_s. \tag{9.3}$$

Summing over r and s, and dividing by n^2 gives

$$\frac{1}{n^2}\sum_{r=1}^{n}\sum_{s=1}^{p}d_{rs}^2 = \frac{2}{n}\sum_{r=1}^{n}\mathbf{x}_r'\mathbf{x}_r. \tag{9.4}$$

Hence, from equations (9.3) and (9.4) we have

$$\mathbf{x}_r'\mathbf{x}_r = \frac{1}{n}\sum_{r=1}^{n}d_{rs}^2 - \frac{1}{2n^2}\sum_{r=1}^{n}\sum_{s=1}^{n}d_{rs}^2,$$

with a similar expression for $\mathbf{x}_s'\mathbf{x}_s$. Writing equation (9.2) as

$$\mathbf{x}_r'\mathbf{x}_s = -\tfrac{1}{2}(d_{rs}^2 - \mathbf{x}_r'\mathbf{x}_r - \mathbf{x}_s'\mathbf{x}_s),$$

and then substituting for $\mathbf{x}_r'\mathbf{x}_r$ and $\mathbf{x}_s'\mathbf{x}_s$ from the previous equations, gives

$$\mathbf{x}_r'\mathbf{x}_s = -\frac{1}{2}\left(d_{rs}^2 - \frac{1}{n}\sum_{r=1}^{n}d_{rs}^2 - \frac{1}{n}\sum_{s=1}^{n}d_{rs}^2 + \frac{1}{n^2}\sum_{r=1}^{n}\sum_{s=1}^{n}d_{rs}^2\right).$$

Let

$$a_{rs} = -\tfrac{1}{2}d_{rs}^2$$

$$a_{r.} = n^{-1}\sum_{s}a_{rs}$$

$$a_{.s} = n^{-1}\sum_{r}a_{rs}$$

$$a_{..} = n^{-2}\sum_{r}\sum_{s}a_{rs},$$

and so

$$\mathbf{x}_r'\mathbf{x}_s = a_{rs} - a_{r.} - a_{.s} + a_{..},$$

which gives us the (r, s)th element of $\mathbf{B}$ in terms of the distances $\{d_{rs}^2\}$.
    Define matrix $\mathbf{A}$ as $[\mathbf{A}]_{rs} = a_{rs}$. Recall that the centring matrix $\mathbf{H}$ is given by

$$\mathbf{H} = \mathbf{I} - n^{-1}\mathbf{1}\mathbf{1}',$$

and then it is easily seen that $\mathbf{B}$ can be obtained from $\mathbf{A}$ as

$$\mathbf{B} = \mathbf{HAH}.$$

To find $\mathbf{x}_r$ We now need to find the coordinates $\mathbf{x}_r$ from the matrix $\mathbf{B}$. Let the coordinates $\mathbf{x}_r$ be placed in a matrix $\mathbf{X}$, so that

$$\mathbf{X} = [\mathbf{x}_1, \dots, \mathbf{x}_n]'.$$

Then

$$\mathbf{B} = \mathbf{XX}'. \tag{9.5}$$

Table 9.2 Matrix A for the French cities

	Bord	Boul	Bour	Caen	Dijo	Lyon	Mars	Nants	Pari	Renn	Roue	Toul
Bordeaux	0	.	.	.	.	.	.	.	.	.	.	.
Boulogne	−1428	0	.	.	.	.	.	.	.	.	.	.
Bourges	−3535	−500	0	.	.	.	.	.	.	.	.	.
Caen	−7202	−151	−304	0	.	.	.	.	.	.	.	.
Dijon	−819	−616	−125	−627	0	.	.	.	.	.	.	.
Lyon	−696	−1125	−180	−925	−97	0	.	.	.	.	.	.
Marseille	−781	−2376	−696	−1901	−605	−224	0	.	.	.	.	.
Nantes	−231	−638	−274	−174	−756	−819	−1462	0	.	.	.	.
Paris	−769	−135	−120	−125	−211	−490	−1328	−361	0	.	.	.
Rennes	−432	−423	−328	−72	−794	−980	−1786	−31	−296	0	.	.
Rouen	−858	−65	−238	−36	−432	−769	−1748	−304	−36	−198	0	.
Toulouse	−135	−2000	−490	−1232	−769	−396	−304	−661	−1066	−952	−1296	0
($\times 10^{-1}$)												

Matrix **B** is symmetric, positive semi-definite and it has rank p, since $r(\mathbf{B}) = r(\mathbf{XX'}) = r(\mathbf{X}) = p$. Thus **B** has p positive eigenvalues and $n − p$ zero eigenvalues. Write **B** in terms of its spectral decomposition,

$$\mathbf{B} = \mathbf{V\Lambda V'},$$

where $\mathbf{\Lambda} = \mathrm{diag}(\lambda_1, \lambda_2, \ldots, \lambda_n)$, the diagonal matrix of the eigenvalues $\{\lambda_i\}$ of **B**, and $\mathbf{V} = [\mathbf{v}_1, \ldots, \mathbf{v}_n]$, the matrix of corresponding eigenvectors, normalized so that $\mathbf{v}'_i \mathbf{v}_i = 1$. The eigenvalues are labelled so that $\lambda_1 \geq \lambda_2 \geq \ldots \geq \lambda_n \geq 0$. Now, **B** has $n − p$ eigenvalues that are zero, and so $\lambda_{p+1} = \lambda_{p+2} = \ldots = \lambda_n = 0$. Exclude these from the spectral decomposition of **B**, and so

$$\mathbf{B} = \mathbf{V}_1 \mathbf{\Lambda}_1 \mathbf{V}'_1,$$

where $\mathbf{\Lambda}_1 = \mathrm{diag}(\lambda_1, \ldots, \lambda_p)$, and $\mathbf{V}_1 = [\mathbf{v}_1, \ldots, \mathbf{v}_p]$.
We now write this equation as

$$\mathbf{B} = [\mathbf{V}_1 \mathbf{\Lambda}_1^{\frac{1}{2}}][\mathbf{V}_1 \mathbf{\Lambda}_1^{\frac{1}{2}}]',$$

where $\mathbf{\Lambda}_1^{\frac{1}{2}} = \mathrm{diag}(\lambda_1^{\frac{1}{2}}, \ldots, \lambda_p^{\frac{1}{2}})$. Comparing this to equation (9.5), we see that

$$\mathbf{X} = \mathbf{V}_1 \mathbf{\Lambda}_1^{\frac{1}{2}}.$$

Thus we have recovered the coordinate matrix **X**.

Applying the above to the distances between the French cities, we first find matrix $\mathbf{A} = [-\frac{1}{2}d_{rs}^2]$, which is shown in Table 9.2, and where elements have been severely approximated for reasons of space. Next, matrix $\mathbf{B} = \mathbf{HAH}$ is found, and is shown in Table 9.3. Next the eigenvalues and eigenvectors of **B** are found as shown in Table 9.4.

The first two eigenvalues are much greater than the other ten, some of which are negative. If the original distances between the cities had been exact (and if the world were flat), and there were no rounding errors in the calculations, all of these ten eigenvalues would have the value zero. Note that there will always be one eigenvalue that has the value zero, whatever the circumstances, since $\mathbf{B1} = \mathbf{HAH1} = \mathbf{0} = 0\mathbf{1}$.

Table 9.3 Matrix **B** for the French cities

	Bord	Boul	Bour	Caen	Dijo	Lyon	Mars	Nants	Pari	Renn	Roue	Toul
Bordeaux	617	.	.	.	.	.	.	.	.	.	.	.
Boulogne	−625	989	.	.	.	.	.	.	.	.	.	.
Bourges	−37	2	14	.	.	.	.	.	.	.	.	.
Caen	−183	572	−68	458	.	.	.	.	.	.	.	.
Dijon	−317	73	76	−204	388	.	.	.	.	.	.	.
Lyon	−123	−366	92	−431	362	530	.	.	.	.	.	.
Marseille	335	−1074	118	−865	−397	848	1615	.	.	.	.	.
Nantes	260	39	−84	237	−380	−372	−472	−1462	.	.	.	.
Paris	−342	477	5	222	101	−107	−4038	−61	236	.	.	.
Rennes	107	303	−90	388	−369	−484	7486	382	52	462	.	.
Rouen	−345	635	−26	397	−33	−299	−736	83	286	237	410	.
Toulouse	655	−1024	−1	−522	−93	350	985	3	−466	−240	−610	963
$(\times 10^{-1})$												

Table 9.4 Eigenvalues and eigenvectors of **B**

E'value	50473	20541	−861	750	−564	284	−152	97	−70	−32	−3	0
E'vec	0.18	−0.47	−0.44	0.56	−0.38	−0.01	−0.04	−0.02	−0.11	0.02	−0.04	0.29
	−0.42	0.26	0.43	0.11	−0.61	0.30	−0.05	−0.07	−0.09	−0.04	−0.04	0.29
	0.03	0.08	0.02	0.17	0.42	0.38	−0.33	0.36	−0.01	−0.15	−0.54	0.29
	−0.29	−0.10	−0.03	−0.18	0.15	−0.42	0.30	−0.23	−0.51	−0.17	−0.39	0.29
	0.08	0.42	−0.10	0.25	0.33	0.06	−0.23	−0.49	−0.25	0.29	0.33	0.29
	0.25	0.32	−0.52	−0.59	−0.31	0.11	−0.08	0.06	0.00	−0.06	−0.05	0.29
	0.55	0.20	0.30	0.14	0.02	0.05	0.67	0.12	0.05	0.01	−0.01	0.29
	−0.11	−0.38	0.08	−0.25	0.18	0.28	0.04	0.36	−0.35	−0.04	0.57	0.29
	−0.17	0.20	−0.08	0.18	0.09	−0.36	−0.05	0.10	0.39	−0.63	0.32	0.29
	−0.22	−0.33	−0.07	−0.17	0.18	0.33	0.23	−0.45	0.56	0.10	−0.07	0.29
	−0.28	0.09	−0.05	0.01	0.01	−0.38	0.02	0.43	0.23	0.67	−0.02	0.29
	0.40	−0.29	0.48	−0.23	−0.08	−0.33	−0.48	−0.17	0.10	0.03	−0.05	0.29

Ignoring the theoretically zero eigenvalues, the coordinate matrix $\mathbf{X} = \mathbf{V}_1 \mathbf{\Lambda}_1^{\frac{1}{2}}$ is given by

$$
\begin{bmatrix}
0.18 & -0.47 \\
-0.42 & 0.26 \\
0.03 & 0.08 \\
-0.29 & -0.10 \\
0.08 & 0.42 \\
0.25 & 0.32 \\
0.55 & 0.20 \\
-0.11 & -0.38 \\
-0.17 & 0.20 \\
-0.22 & -0.33 \\
-0.28 & 0.09 \\
0.40 & -0.29
\end{bmatrix}
\begin{bmatrix}
224.7 & 0 \\
0 & 143.3
\end{bmatrix}
=
\begin{bmatrix}
41.0 & -67.1 \\
-94.2 & 36.6 \\
6.3 & 11.9 \\
-66.0 & -13.9 \\
17.5 & 59.9 \\
57.0 & 45.5 \\
124.3 & 28.8 \\
-25.0 & -54.6 \\
-39.2 & 28.1 \\
-49.1 & -46.8 \\
-62.3 & 13.1 \\
89.7 & -41.7
\end{bmatrix}
$$

Figure 9.1 shows a map of the cities plotted from these coordinates. First, we notice that the map is not in the usual orientation of a map of France. The map would need to be rotated approximately 90° clockwise, in order to give the usual orientation.

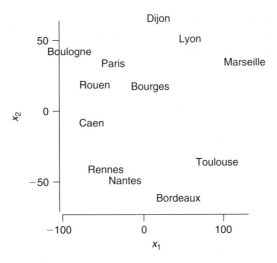

Figure 9.1 Reconstruction of a map of French cities

9.1.2 Classical scaling

Classical scaling is not about reproducing maps of France or other countries. It is about producing maps (configurations) of points, where each point represents an individual or object, and the distance between a pair of points represents the dissimilarity measured between the corresponding individuals or objects. To do this, we simply replace the distances, $\{d_{rs}\}$, between cities in the above algorithm, with dissimilarities, $\{\delta_{rs}\}$, between individuals or objects.

It can be shown that when $\mathbf{A}$, and then $\mathbf{B}$, are formed from a set of dissimilarities, if $\mathbf{B}$ is positive semi-definite of rank p, then a configuration of points can be found in p dimensions such that d_{rs}, the distance between points r and s, is equal to the original dissimilarity, δ_{rs}, between the corresponding objects r and s for all r and s. However, in practice, the rank of $\mathbf{B}$ will be large and usually equal to $n - 1$. Finding a space of $n - 1$ dimensions in which to represent the data is hardly dimension reduction. However, we could rotate the configuration of points to its principal axes, where the projected points onto the first principal axis have maximum variation, the projected points onto the second principal axis have maximum variation, but where the second principal axis is orthogonal to the first, and so on. Then the first p principal axes are chosen to represent the configuration, where p is usually chosen as two or three. Now

$$\mathbf{X}'\mathbf{X} = (\mathbf{V}_1\mathbf{\Lambda}_1^{\frac{1}{2}})'(\mathbf{V}_1\mathbf{\Lambda}_1^{\frac{1}{2}}) = \mathbf{\Lambda}_1^{\frac{1}{2}}\mathbf{V}_1'\mathbf{V}_1\mathbf{\Lambda}_1^{\frac{1}{2}'} = \mathbf{\Lambda},$$

and since $\mathbf{\Lambda}$ is a diagonal matrix, with the diagonal elements decreasing down the diagonal, it means that $\mathbf{X}$ is already referred to its principal axes, and nothing else has to be done. We simply choose the first two or three dimensions of $\mathbf{X}$. Because $\mathbf{X}$ is referred to its principal axes, another name for classical scaling is 'principal coordinates analysis' (PCO).

Another potential problem is that when $\mathbf{B}$ is not positive semi-definite, negative eigenvalues occur. To overcome this, a positive constant can be added to all the dissimilarities (but not δ_{rr}) to make $\mathbf{B}$ positive semi-definite. Finding the smallest value needed to ensure this is called the *additive constant problem*. However, it is more common in practice to simply ignore the negative eigenvalues, and carry on regardless.

Now, if **B** is positive semi-definite

$$\frac{1}{2}\sum_{r=1}^{n}\sum_{s=1}^{n}d_{rs}^2 = n\sum_{r=1}^{n}\mathbf{x}_r'\mathbf{x}_r = n\text{tr }\mathbf{B} = n\sum_{r=1}^{n}\lambda_r.$$

A measure of the proportion of variation explained by using only the first p dimensions is

$$\frac{\sum_{i=1}^{p}\lambda_i}{\sum_{i=1}^{n-1}\lambda_i}.$$

If **B** is not positive semi-definite, we can use

$$\frac{\sum_{i=1}^{p}\lambda_i}{\sum_{i=1}^{n-1}|\lambda_i|} \quad \text{or} \quad \frac{\sum_{i=1}^{p}\lambda_i}{\sum(\text{positive eigenvalues})}.$$

9.1.3 Classical scaling of golfers

The scores for the first round of the Open Golf Championship, used in Chapter 5 for principal components analysis, were subjected to a classical scaling analysis. The city block metric was used to calculate dissimilarities between pairs of golfers. Figure 9.2 shows the configuration of points representing the golfers, using the first two principal axes.

This configuration is somewhat similar to the one in Figure 5.6, the principal component scores for the golfers. Indeed, if the covariance matrix had been used for the PCA, and Euclidean distance for dissimilarities in classical scaling, then the configurations would

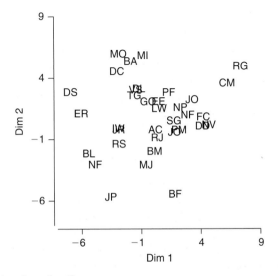

Figure 9.2 Classical scaling of golfers

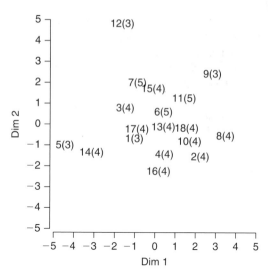

Figure 9.3 Classical scaling of holes

have been identical. This is because classical scaling and PCA are equivalent under these conditions (see Cox and Cox, 2000).

Figure 9.3 shows the results of a classical scaling analysis for the holes, where the roles of golfers and holes have been reversed. Also shown in the plot labels are the par scores for the holes. So, for instance, 14(4) represents hole 14 which is a par 4 hole. Again, the city block metric was used to find dissimilarities between pairs of holes, but with each score for a hole standardized by dividing by the mean score for that hole. This was done to place the holes on an equal footing, otherwise the resulting configuration simply clusters the par 3 holes together, the par 4 holes together, and so on. The first two eigenvalues in the analysis were 68.5 and 55.3. The sum of the eigenvalues was 355.5, and so the first two dimensions account for only 35% of the 'variation' explained. Four dimensions would have explained 60%.

From the plot, it can be seen that the par 4 holes do tend to cluster together, holes 5 and 14 are close together away from the others, hole 12 is on its own away from the others, and the par 5 holes are close to each other. Also note that the group of holes identified in PCA, $\{3, 4, 5, 14, 16, 17\}$, are all to the lower left of the plot.

9.1.4 Sammon maps

Classical scaling is also called metric scaling, as opposed to nonmetric scaling discussed in the next section. Other forms of metric scaling exist, 'Sammon maps' being one of them (Sammon, 1969). A configuration is found such that the distances, $\{d_{rs}\}$, between pairs of points, minimize the loss function, S, where

$$S = \frac{\sum_{r<s} w_{rs}(d_{rs} - \delta_{rs})^2}{\sum_{r<s} \delta_{rs}},$$

and w_{rs} is a weight function. If w_{rs} is chosen as δ_{rs}^{-1}, then more weight is given to small dissimilarities, while if w_{rs} is chosen as δ_{rs}, more weight is given to larger dissimilarities.

Of course, the choice of w_{rs} is subjective, but usually smaller dissimilarities are deemed to be more important than larger ones. There are other possible variations on the loss function.

The distances, $\{d_{rs}\}$, in the loss function are written in terms of their coordinates, $d_{rs}^2 = \sum_{i=1}^{p} (x_{ri} - x_{si})^2$, and then S is minimized by using the method of steepest descent. If x_{ri}^m is the value for x_{ri} in the mth iteration in minimizing S, then

$$x_{ri}^{(m+1)} = x_{ri}^{(m)} - h \frac{\partial S}{\partial x_{ri}} \Big/ \left| \frac{\partial^2 S}{\partial x_{ri}^2} \right|,$$

where h is an appropriate step length.

9.2 Nonmetric scaling

Like classical scaling, nonmetric scaling finds a configuration of points representing objects or individuals, usually in two-dimensional Euclidean space. The difference between non-metric scaling and classical scaling is that the magnitude of the distances between pairs of points in the space no longer approximate the corresponding magnitude of original dissimilarities, but the rank order of the distances matches the rank order of the dissimilarities (as well as possible). The matching rank orders is the 'non-metric' idea behind nonmetric scaling.

Consider a configuration of n points in a two-dimensional space, representing our objects, with distances between all pairs of points, $\{d_{rs}\}$. A measure of how well the configuration of points represents the dissimilarities is given by the loss function, S, where

$$S = \sqrt{\frac{S^*}{T^*}},$$

(9.6)

$$S^* = \sum_{r<s}^{n} (d_{rs} - \hat{d}_{rs})^2, \quad T^* = \sum_{r<s} d_{rs}^2,$$

and $\{\hat{d}_{rs}\}$ are 'disparities', given by the primary monotone least squares regression of $\{d_{rs}\}$ on $\{\delta_{rs}\}$. These disparities need further explanation which we give using the following example.

The dissimilarities between six objects are given in the matrix below, and are displayed in Table 9.5 in rank order:

$$[\delta_{rs}] = \begin{bmatrix} 0.0 & 0.7 & 1.3 & 1.5 & 0.5 & 1.1 \\ 0.7 & 0.0 & 0.2 & 1.0 & 0.1 & 1.4 \\ 1.3 & 0.2 & 0.0 & 0.8 & 0.3 & 1.8 \\ 1.5 & 1.0 & 0.8 & 0.0 & 0.9 & 1.7 \\ 0.5 & 0.1 & 0.3 & 0.9 & 0.0 & 1.4 \\ 1.1 & 1.4 & 1.8 & 1.7 & 1.4 & 0.0 \end{bmatrix}.$$

The points in the configuration in Figure 9.4 represent these objects, the distances between the points also being displayed in Table 9.5, according to the rank order of the dissimilarities. If the rank order of the distances were exactly the same as the rank order of the dissimilarities,

Table 9.5 Ordered dissimilarities, distances and disparities

δ_{rs}:	δ_{25}	δ_{23}	δ_{35}	δ_{15}	δ_{12}	δ_{34}	δ_{24}	δ_{45}	δ_{16}	δ_{13}	δ_{56}	δ_{26}	δ_{14}	δ_{46}	δ_{36}
	0.1	0.2	0.3	0.5	0.7	0.8	0.9	1.0	1.1	1.3	1.4	1.4	1.5	1.7	1.8
d_{rs}:	0.9	2.7	3.4	3.3	3.0	3.8	4.2	4.5	5.9	5.2	6.1	6.8	7.4	9.3	9.4
$\hat{d}_{rs}$:	0.9	2.7	3.23	3.23	3.23	3.8	4.2	4.5	5.55	5.55	6.45	6.45	7.4	9.3	9.4

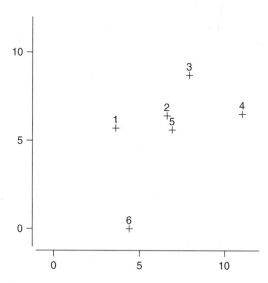

Figure 9.4 Configuration of points representing the objects

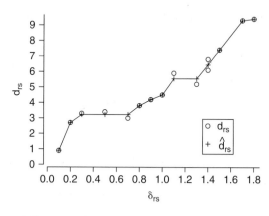

Figure 9.5 Distances and disparities plotted against dissimilarities

then the configuration of points would be deemed to be a perfect representation of the objects. In practice, the two rank orders will not usually match. Figure 9.5 shows the distances (circles) plotted against the corresponding dissimilarities. The monotone least squares regression of the distances on the dissimilarities is given by the crosses, which are joined together for visual purposes. To find this regression, imagine starting at the bottom left of the plot of distances against dissimilarities, and drawing in the crosses, working towards the top right. The crosses represent the values of the disparities, read from the y-axis, on the same scale as the distances. At each step you are trying to place a cross on top of a circle, so that the disparity $\hat{d}_{rs}$ is equal to the distance d_{rs}, so that the contribution $(d_{rs} - \hat{d}_{rs})^2$ to the loss function in (9.6) is zero. However, as you plot the crosses, you are not allowed to decrease from a value previously plotted. Thus for the 3rd, 4th and 5th points, the mean of the three distances associated with these points is found, and then a cross is placed at this value for each of the three points.

Similarly for the 7th and 8th points, and also the 9th and 10th. For these cases there will be a positive contribution to the loss function from $(d_{rs} - \hat{d}_{rs})^2$. Defining the disparities in this way minimizes the numerator, $S^* = \sum_{r<s}^{n} (d_{rs} - \hat{d}_{rs})^2$, of the loss function, with respect to $\{\hat{d}_{rs}\}$, for given $\{d_{rs}\}$. This gives the value of the loss function for this particular configuration, noting that the denominator in the loss function standardizes the function. A configuration is now sought that has a minimum value for the loss function, which we will now call STRESS, as introduced by Kruskal.

Finding a configuration that minimizes S is not an easy task, and has to be done iteratively. An algorithm for doing so is:

1. Choose an initial configuration (usually arbitrary).
2. Normalize the configuration to have its centroid at the origin and unit mean square distance from the origin.
3. Find $\{d_{rs}\}$ from the normalized configuration.
4. Fit $\{\hat{d}_{rs}\}$ and calculate STRESS S.
5. The STRESS, S, is considered as a function of the coordinates defining the points, i.e. the distances in the definition of S are replaced by the function defining them; for example, for Euclidean distance in two dimensions, $d_{rs} = ((x_{r1} - x_{s1})^2 + (x_{r2} - x_{s2})^2)^{1/2}$. Let all the $2 \times n$ coordinates be placed in a vector $\mathbf{x}$. The gradient $\partial S / \partial \mathbf{x}$ is then found.

   If $|\partial S / \partial \mathbf{x}| < \varepsilon$, where ε is a preselected very small number, then a configuration with a minimum stress has been found. Note this could be a local minimum and not the global minimum.
6. Move to a new configuration, along the path of steepest descent, i.e.

$$\mathbf{x}_{\text{new}} = \mathbf{x}_{\text{old}} - l \frac{\partial S}{\partial \mathbf{x}} \bigg/ \left| \frac{\partial S}{\partial \mathbf{x}} \right|$$

   where l is an appropriate step length.
7. Go to step 2.

To overcome the problem of reaching a local minimum, it is recommended that several different starting configurations are used. The final value of STRESS is an indicator of how well the rank order of the distances matches the rank order of the original dissimilarities.

9.3 Crime rates in US states

One of the SAS datasets used for illustration consists of the crime rates (*murder, rape, robbery, assault, burglary, larceny, car theft*) per 100 thousand people for each state in America. These rates were standardized so that each rate had a range [0, 1], and then the data were subjected to nonmetric MDS, using Euclidean distance. The final STRESS was 9%.

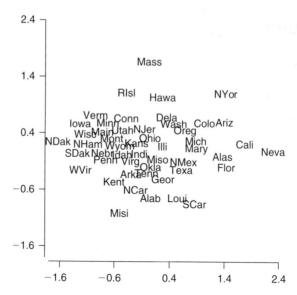

Figure 9.6 Nonmetric MDS of crime

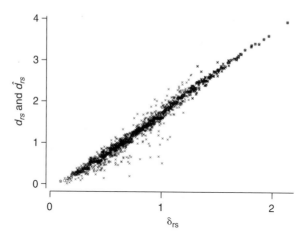

Figure 9.7 Shepard plot for crime

Figure 9.6 shows the configuration obtained, with states close to each other having similar crime profiles. States to the left of the plot have least crime and states to the right, the most. North Dakota, South Dakota and West Virginia have the least crime; Nevada, California and New York have the most. Massachusetts at the top of the plot has the most car theft, Mississippi at the bottom, has the least.

Figure 9.7 shows the Shepard plot for the crime data. This is a plot of the monotone least squares regression of $\{d_{rs}\}$ on $\{\delta_{rs}\}$. However, since there are so many points to plot ($\frac{1}{2}n(n-1)$ in total), it is difficult to see what is happening in fine detail. The crosses represent the values of d_{rs} plotted against δ_{rs}; the circles represent the values of $\hat{d}_{rs}$ plotted against δ_{rs}. In general, as the points are reasonably tight to the regression, the model is sound.

9.4 Procrustes analysis

Suppose we have two configurations of points representing the same set of objects or individuals. For instance, one configuration may have arisen from classical scaling and the other from nonmetric scaling. How can we match one configuration to the other by translation, rotation and reflection? The technique for doing this is Procrustes analysis. In Greek mythology, Damastes was an innkeeper on the road from Eleusis to Athens. He had a peculiar way of making sure his guests were comfortable and fitted into his beds. If his guests were too long, he would chop their legs off to make them fit. If his guests were too short, he would stretch them on a rack to make them fit. Damastes was given the nickname Procrustes, which means 'stretcher'. Sibson (1978) gives a review of Procrustes analysis, where the solution of finding the optimal translation, rotation and reflection is given. Here we simply state the results.

Let the coordinates of a configuration of n points in q-dimensional Euclidean space be placed in an $n \times q$ matrix $\mathbf{X}$. Suppose this needs to be matched to another configuration of points in a p-dimensional Euclidean space, with coordinates placed in the $n \times p$ matrix $\mathbf{Y}$ $(p \geq q)$. Note that we are assuming that the rth point in the first configuration is in a one-to-one correspondence with the rth point in the second configuration.

Firstly, $p - q$ columns of zeros are placed at the end of matrix $\mathbf{X}$ in order to give the two configurations the same number of dimensions, and note that in this case, the dimension of the configuration to be matched is less than or equal to the dimension of the configuration to be matched to – it cannot be the other way round.

The sum of squared distances between the points in the $\mathbf{Y}$ space and the corresponding points in the $\mathbf{X}$ space is

$$R^2 = \sum_{r=1}^{n} (\mathbf{y}_r - \mathbf{x}_r)'(\mathbf{y}_r - \mathbf{x}_r),$$

where $\mathbf{x}_r$ and $\mathbf{y}_r$ are the coordinates of the rth point in each of the two configurations, i.e. the rth rows of $\mathbf{X}$ and $\mathbf{Y}$.

Translation, dilation and rotation of $\mathbf{X}$ is given by

$$\mathbf{y}_r = \rho \mathbf{A}'\mathbf{x}_r + \mathbf{b}.$$

The translation is given by $\mathbf{b}$, a constant vector, the dilation by ρ, a constant, and the rotation by $\mathbf{A}$, an orthogonal matrix.

The sum of squared distances, R^2, is minimized, where it can be shown that the *optimal translation* is to translate the centroid of $\mathbf{X}$ to the centroid of $\mathbf{Y}$. We now assume that this has been done by translating both configurations to have centroid at the origin, i.e.

$$\sum_{r=1}^{n} \mathbf{x}_r = \sum_{r=1}^{n} \mathbf{y}_r = \mathbf{0}.$$

The *optimal dilation* is given by

$$\rho = \text{tr}(\mathbf{X}'\mathbf{YY}'\mathbf{X})^{1/2}/\text{tr}(\mathbf{X}'\mathbf{X}).$$

The *optimal rotation* is given by

$$\mathbf{A} = (\mathbf{X}'\mathbf{YY}'\mathbf{X})^{1/2}(\mathbf{Y}'\mathbf{X})^{-1}.$$

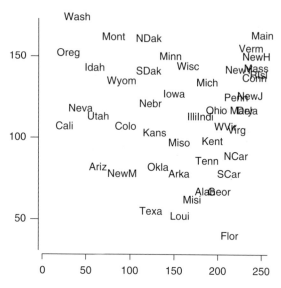

Figure 9.8 American states

Substituting the optimal values back into R^2 gives its minimum as

$$R^2 = 1 - \{\text{tr}(\mathbf{X}'\mathbf{Y}\mathbf{Y}'\mathbf{X})^{1/2}\}^2 / \{\text{tr}(\mathbf{X}'\mathbf{X})\text{tr}(\mathbf{Y}'\mathbf{Y})\},$$

which can be used to assess the matching of the configurations, and is called the Procrustes statistic. It can be shown that $0 \leq R^2 \leq 1$.

Example

As an example of Procrustes analysis, the configuration of points representing the states of America in the crime example are matched to the actual position of the states, geographically. A crude estimate of the centre of each state was determined by eye from a map of the USA. The map of states from the crime data was then matched to the geographical map of the states. Figure 9.8 shows a map of the states geographically. Figure 9.9 shows a map of the states, crime-wise, matched to the geographical map. The value of the Procrustes statistic was 0.83, which is rather high, implying that there is not a very close match of the two configurations of the states. However, perusal of the positions of the individual states does show there to be some agreement. For example, Kentucky, California, and Tennessee lie close to their geographical positions, while Massachusetts, New Jersey, and New York do not. The impression is that the states in the north west have been pushed towards the centre of the country. If the Procrustes match is accepted, then there is a geographical pattern to crime.

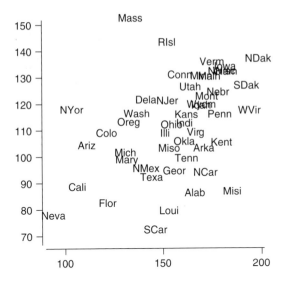

Figure 9.9 Crime configuration matched to the USA map

9.5 Individual differences scaling

If we have a set of dissimilarities $\{\delta_{rs}\}$ for objects, measured for each of N individuals, then these could be analysed separately for each individual, or by 'Individual Differences Scaling', which analyses the data as a whole. Carroll and Chang (1970) proposed the INDSCAL model (the acronym stands for, **IN**dividual **D**ifferences **SCAL**ing). The model comprises two spaces. Points in the 'group stimulus' space form an underlying or common configuration for the objects, and points in a subjects space represent the individuals. Both spaces are of dimension p, usually chosen as two. Let the points in the group stimulus space have coordinates x_{rt} $(r = 1, \ldots, n; t = 1, \ldots, p)$. Let the points in the individuals space have coordinates w_{it} $(i = 1, \ldots, N; t = 1, \ldots, p)$, and these are viewed as 'weights' for the ith individual. Let the dissimilarities for the ith individual be denoted by $\{\delta_{rs,i}\}$. The weighted Euclidean distance between the rth and sth objects for the ith individual is

$$d_{rs,i} = \left\{ \sum_{t=1}^{p} w_{it}(x_{rt} - x_{st})^2 \right\}^{\frac{1}{2}}.$$

The individual weights $\{w_{it}\}$ and object coordinates $\{x_{rt}\}$ are found that match $\{d_{rs,i}\}$ to $\{\delta_{rs,i}\}$. For each individual, a cross products matrix, $\mathbf{B}_i$, is found from the dissimilarities, where

$$\mathbf{B}_i = \left[\sum_{t=1}^{p} w_{it} x_{rt} x_{st} \right] = \mathbf{H} \mathbf{A}_i \mathbf{H},$$

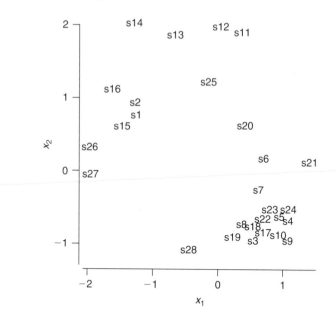

Figure 9.10 Group stimulus space for occlusal surfaces

as for classical scaling. Then least squares estimates of $\{w_{it}\}$ and $\{x_{rt}\}$ are found by minimizing

$$ S = \sum_{r,s,i} \left(b_{rs,i} - \sum_{t=1}^{p} w_{it} x_{rt} x_{st} \right)^2. $$

Once the solution has been found, the group stimulus space gives an 'average' configuration of points representing the objects. The individuals space plots the weights for each individual. If the coordinates of the points in the group stimulus space are weighted by the weights for the ith individual, and then replotted, the resulting plot is a configuration of points representing the objects for that ith individual.

9.5.1 Individual differences for rats

Andrews and Herzberg (1985) have a set of data relating to caries on the occlusal surfaces of rats' teeth, for eight groups of rats fed different diets. For each of the 28 occlusal surfaces, caries was scored as 0, 1 or 2. Interest lies in the occlusal surfaces, labelled S1 to S28, and not on the rats themselves. Thirteen rats were chosen from each group and dissimilarity between occlusal surfaces was based on the city block metric, taken over the thirteen rats. The diets formed the 'subjects', and the occlusal surfaces, the 'stimuli'. The analysis was carried out using the SAS version of INDSCAL in PROC MDS.

Figure 9.10 shows the group stimulus space, and Figure 9.11, the subjects space. Looking at the raw data, it can be seen that the occlusal surfaces in the group at the lower right of the group stimulus space have a much higher rate of caries than the other surfaces. Surface 15 has the lowest rate, while surfaces 9 and 10 have the highest rates. For the diets, diet 1 has

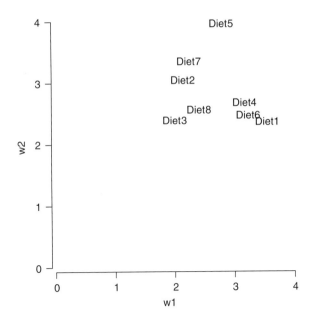

Figure 9.11 Subjects space for diets

by far the lowest caries rate, then diet 5, while the rest have very similar rates. Plots of the occlusal surfaces for the individual diets have not been included here for reasons of space. It can be seen that diet 1 would stretch the x_1 axis, diet 5 would stretch the x_2 axis, diet 3 would give a configuration similar to that of the group stimulus space, and the other diets would do something in between these.

9.6 Exercises

1. Let the Euclidean distances between three points, P_1, P_2 and P_3, be

$$d_{12} = \sqrt{10}, \quad d_{23} = \sqrt{5}, \quad d_{23} = \sqrt{13}.$$

Using the theory of Section 9.1, show that

$$\mathbf{B} = \frac{1}{9} \begin{pmatrix} 41 & -16 & -25 \\ -16 & 17 & -1 \\ -25 & -1 & 26 \end{pmatrix}.$$

Find the eigenvalues and eigenvectors of **B** and hence recover coordinates for P_1, P_2 and P_3. Confirm that the distances between these points are the same as the original distances.

2. The table below gives the percentage number of candidates gaining the various grades in British public examinations (GCSE), for twenty-four subjects in 1991. Carry out classical scaling (or nonmetric scaling) on the subjects using Euclidean distance as a

measure of dissimilarity. Represent the subjects as points in a two-dimensional space. Repeat the analysis using the city block metric as the measure of dissimilarity.

	A	B	C	D	E	F	G
Art and Design	13.2	15.0	21.9	19.1	15.2	10.3	4.7
Business Studies	9.7	12.1	24.5	18.8	16.4	10.8	5.5
Latin	52.3	23.1	12.9	6.0	2.9	1.5	0.8
Greek	61.0	20.1	9.7	3.3	1.5	1.6	1.0
Classical Civilization	21.5	26.6	21.1	10.8	8.3	5.2	4.0
Computer Studies	9.0	13.8	21.2	19.0	15.1	10.7	7.2
CDT	8.7	13.1	17.9	18.7	18.1	13.7	7.2
Economics	11.0	18.6	23.8	16.8	13.6	9.2	4.6
English	9.4	17.2	27.1	22.0	14.4	7.4	2.4
English Literature	11.6	19.5	26.6	20.9	13.3	6.2	1.7
Home Economics	6.2	12.5	20.2	20.6	18.7	13.6	7.5
Geography	12.7	15.8	19.4	18.2	14.8	10.9	5.7
History	13.5	18.2	20.2	16.2	13.1	10.3	6.0
French	17.9	13.0	15.2	16.7	13.5	14.1	8.4
German	22.6	17.1	18.3	16.0	10.7	9.1	5.6
Spanish	29.3	17.6	14.1	13.0	9.4	9.9	5.9
Mathematics	8.0	9.6	25.7	16.9	17.1	12.6	5.6
Music	17.9	21.6	22.4	14.1	10.5	7.8	4.5
Religious Studies	12.3	17.3	21.2	15.5	13.2	10.1	7.5
Biology	13.3	15.2	21.4	17.5	13.9	10.3	5.6
Chemistry	16.4	20.1	21.6	15.7	11.8	8.3	4.3
Physics	15.9	16.4	25.1	18.0	11.5	7.9	3.7
Science	7.0	10.3	19.5	19.5	17.6	13.3	7.7
Social Sciences	4.9	12.2	20.5	17.1	17.3	14.5	9.5

3. Use the Jaccard coefficient to find a similarity matrix from the whisky data described in Chapter 8, and found in Cox and Cox (2000). Transform this into a dissimilarity matrix and then carry out a non-metric scaling analysis.

4. Following on from Exercise 1, let Q_1, Q_2 and Q_3 be three points in a Euclidean space, with coordinates $(-0.72, 1.74)$, $(1.02, 4.37)$ and $(2.70, 2.89)$, where Q_i corresponds to P_i. Use Procrustes analysis to match the configuration of points Q_1, Q_2, Q_3, to the configuration of points, P_1, P_2, P_3.

5. Following on from Exercise 2, the corresponding percentage grades for 1992 were

	A	B	C	D	E	F	G
Art and Design	13.9	15.6	22.6	19.2	14.5	9.6	4.0
Business studies	12.5	13.6	20.9	19.5	16.4	11.3	4.7
Latin	61.0	20.6	10.7	4.2	2.0	0.9	0.5
Greek	71.1	15.3	8.1	3.3	1.5	0.2	0.4
Classical Civilization	23.6	26.5	21.3	10.8	7.4	4.4	4.7
Computer Studies	9.5	13.8	21.9	17.9	14.8	11.6	7.2
CDT	9.1	13.4	18.4	19.3	18.2	13.0	6.4
Economics	11.8	18.7	24.2	18.1	12.9	8.3	4.0
English	9.6	17.7	28.0	22.0	13.5	6.7	2.1
English Literature	12.0	19.9	27.2	20.1	12.8	5.9	1.6
Home Economics	7.3	13.4	19.9	20.7	18.2	12.3	6.1
Geography	13.5	16.6	20.1	17.8	14.4	10.2	5.5
History	14.0	18.6	20.8	16.6	12.6	9.7	5.4

(Contd)

	A	B	C	D	E	F	G
French	19.1	12.9	15.0	16.5	13.2	13.2	8.4
German	22.8	16.2	16.8	15.8	11.3	10.3	5.9
Spanish	29.7	16.1	14.3	12.7	10.2	10.8	5.3
Mathematics	9.0	10.5	25.9	17.7	17.5	11.7	4.8
Music	20.9	21.9	21.4	14.0	10.3	7.2	3.6
Religious Studies	14.2	18.0	20.7	15.1	12.9	10.0	6.3
Biology	15.5	16.8	25.4	17.4	11.6	7.5	3.8
Chemistry	18.9	23.4	25.4	14.3	9.3	5.2	2.7
Physics	21.5	19.5	25.7	15.4	9.4	5.3	2.4
Science	9.4	12.2	22.3	20.5	16.5	11.2	5.6
Social Sciences	7.5	13.7	22.1	17.8	16.4	11.9	7.6

Carry out classical scaling (or nonmetric scaling) on the subjects, and then match the configuration obtained to that for Exercise 2 using Procrustes analysis.

10

Linear regression analysis

It is assumed that the reader is familiar with *simple linear regression* ('fitting straight lines to data') where there is one response variable, Y, and one explanatory variable, or regressor, X. This chapter first extends simple linear regression to *multiple linear regression*, which caters for more than one explanatory variable. This is then extended to *multivariate multiple linear regression*, which allows for more than one response variable. Books on linear regression analysis include Montgomery and Peck (1992), Kutner et al. (2003) and Weisberg (1985).

10.1 Multiple linear regression

We start with an introductory example using data on the price obtained at auction for thirty-two grandfather clocks. The data are on the OzDASL website (http://www.statsci.org/data/general/auction.html). The data consist of the age of the clock, X_1, the number of bidders for the clock, X_2, and the price paid by the highest bidder, Y. The regression model is

$$y_i = \beta_0 + \beta_1 x_{i1} + \beta_2 x_{i2} + \varepsilon_i \qquad (i = 1, \ldots, 32),$$

where ε_i is the error term, and it is usually assumed that the ε_i's are independent, and have a normal distribution with variance σ^2. To fit the model, we have to estimate $\beta_0, \beta_1, \beta_2$, and σ^2. However, to keep things general at this stage, we consider the general model with q regressor variables

$$y_i = \beta_0 + \beta_1 x_{i1} + \ldots + \beta_q x_{iq} + \varepsilon_i \qquad (i = 1, \ldots, n).$$

Note that the regression model is linear because it is linear in the parameters, $\beta_0, \beta_1, \ldots, \beta_q$. Linearity here does not relate to the x-variables, and so models such as

$$y_i = \beta_0 + \beta_1 x_i + \beta_2 x_i^2 + \beta_3 x_i^3 + \varepsilon_i$$
$$y_i = \beta_0 + \beta_1 \sin(x_{i1}) + \beta_2 \log(x_{i2}) + \varepsilon_i$$

are still linear regression models.

Write the equations for all the observations together,

$$y_1 = \beta_0 + \beta_1 x_{11} + \beta_2 x_{12} + \ldots + \beta_q x_{1q} + \varepsilon_1$$
$$y_2 = \beta_0 + \beta_1 x_{21} + \beta_2 x_{22} + \ldots + \beta_q x_{2q} + \varepsilon_2$$
$$\vdots \quad \vdots \qquad\qquad\qquad\qquad \vdots$$
$$y_n = \beta_0 + \beta_1 x_{n1} + \beta_2 x_{n2} + \cdots + \beta_q x_{nq} + \varepsilon_n.$$

Write this in matrix form

$$
\begin{bmatrix} y_1 \\ y_2 \\ \vdots \\ y_n \end{bmatrix} =
\begin{bmatrix}
1 & x_{11} & x_{12} & \cdots & x_{1q} \\
1 & x_{21} & x_{22} & \cdots & x_{2q} \\
\vdots & & & & \vdots \\
1 & x_{n1} & x_{n2} & \cdots & x_{nq}
\end{bmatrix}
\begin{bmatrix} \beta_0 \\ \beta_1 \\ \vdots \\ \beta_q \end{bmatrix} +
\begin{bmatrix} \varepsilon_1 \\ \varepsilon_2 \\ \vdots \\ \varepsilon_n \end{bmatrix},
$$

i.e.

$$\mathbf{y} = \mathbf{X}\boldsymbol{\beta} + \boldsymbol{\varepsilon}, \tag{10.1}$$

where $\mathbf{y}$ is the vector of y_i's (the *response vector*), $\mathbf{X}$ is the matrix of the explanatory variables, x_{ij}'s, $\boldsymbol{\beta}$ is the vector of β_i's (the *parameter vector*), and $\boldsymbol{\varepsilon}$ is the vector of ε_i's (the *error vector*). Sometimes $\mathbf{X}$ is called the *design matrix*, when it consists of zeros and ones for indicator variables for analysis of variance and covariance models.

Let $\mathbf{x}_i$ be the vector of values of the explanatory variables for the ith observation, preceded by a one. Thus we can write $\mathbf{X}$ in terms of these vectors,

$$
\mathbf{X} = \begin{bmatrix} \mathbf{x}_1' \\ \mathbf{x}_2' \\ \vdots \\ \mathbf{x}_n' \end{bmatrix}.
$$

10.1.1 Least squares estimate of β

Let $\hat{\boldsymbol{\beta}}$ be an estimate of $\boldsymbol{\beta}$. Then the *fitted values*, $\hat{y}_i$, are given by

$$\hat{y}_i = \hat{\beta}_0 + \hat{\beta}_1 x_{i1} + \cdots + \hat{\beta}_q x_{iq} = \mathbf{x}_i'\hat{\boldsymbol{\beta}}.$$

The *residuals* are

$$e_i = y_i - \hat{y}_i.$$

Place the fitted values in a vector, $\hat{\mathbf{y}}$, and the residuals in a vector, $\mathbf{e}$,

$$
\hat{\mathbf{y}} = \begin{bmatrix} \hat{y}_1 \\ \hat{y}_2 \\ \vdots \\ \hat{y}_n \end{bmatrix}, \qquad
\mathbf{e} = \begin{bmatrix} e_1 \\ e_2 \\ \vdots \\ e_n \end{bmatrix},
$$

and then

$$e = y - \hat{y}.$$

The least squares estimate, $\hat{\beta}$, of β, is the value of β that minimizes the sum of squared residuals, SS_E, where

$$SS_E = \sum_{i=1}^{n} e_i^2 = e'e$$

$$= (y - \hat{y})'(y - \hat{y})$$

$$= (y - X\hat{\beta})'(y - X\hat{\beta})$$

$$= y'y - 2\hat{\beta}'X'y + \hat{\beta}'X'X\hat{\beta}.$$

Dropping the hat on β, we have

$$\frac{\partial SS_E}{\partial \beta} = -2X'y + 2X'X\beta,$$

and equating $\partial SS_E / \partial \beta$ to $\mathbf{0}$, gives the least squares estimate, $\hat{\beta}$, of β, as

$$\hat{\beta} = (X'X)^{-1}X'y,$$

provided the inverse, $(X'X)^{-1}$ exists. If the columns of X are linearly independent, then the inverse will exist. It can be shown that $\hat{\beta}$ gives a maximum and not a minimum.

The vector of fitted values is

$$\hat{y} = X\hat{\beta} = X(X'X)^{-1}X'y = Hy,$$

where $H = X(X'X)^{-1}X$ is the *hat matrix*. The hat matrix is not to be confused with the centring matrix, which is also usually denoted by H. The hat matrix converts y to $\hat{y}$, i.e. puts 'hats' on the y's.

The mean vector of $\hat{\beta}$ is given by

$$E(\hat{\beta}) = E((X'X)^{-1}X'y) = (X'X)^{-1}X'E(y) = (X'X)^{-1}X'X\beta = \beta,$$

and hence is unbiased. The variance matrix of $\hat{\beta}$ is given by

$$\text{var}(\hat{\beta}) = \text{var}((X'X)^{-1}X'y) = (X'X)^{-1}X'(\sigma^2 I)((X'X)^{-1}X')' = \sigma^2(X'X)^{-1}.$$

As $\hat{\beta}$ is a linear combination of the y_i's, it has a multivariate normal distribution.

For the residuals,

$$E(e) = \mathbf{0} \qquad \text{var}(e) = \sigma^2(I - H), \tag{10.2}$$

and for the fitted values,

$$E(\hat{y}) = \sigma^2 H.$$

The error variance, σ^2, is estimated by $s^2 = SS_E/(n - q - 1)$.

Example

For the grandfather clock data, the response vector and matrix of explanatory variables are

$$
\mathbf{y} = \begin{bmatrix} 1235 \\ 1080 \\ \vdots \\ 1356 \\ 1262 \end{bmatrix}, \quad
\mathbf{X} = \begin{bmatrix} 1 & 127 & 13 \\ 1 & 115 & 12 \\ \vdots & \vdots & \vdots \\ 1 & 194 & 5 \\ 1 & 168 & 7 \end{bmatrix}.
$$

Then $\mathbf{X'X}$, $(\mathbf{X'X})^{-1}$, and $\mathbf{X'y}$ are given by

$$
\mathbf{X'X} = \begin{bmatrix} 32 & 4638 & 305 \\ 4638 & 695486 & 43594 \\ 305 & 43594 & 3157 \end{bmatrix},
$$

$$
(\mathbf{X'X})^{-1} = \begin{bmatrix} 169545 & -773 & -5705 \\ -773 & 5 & 11 \\ -5705 & 11 & 428 \end{bmatrix} \times 10^{-5},
$$

$$
\mathbf{X'y} = \begin{bmatrix} 42469 \\ 6399156 \\ 418440 \end{bmatrix}.
$$

Hence $\hat{\boldsymbol{\beta}}$ is given by

$$
\hat{\boldsymbol{\beta}} = (\mathbf{X'X})^{-1}\mathbf{X'y} = (-1336.72, 12.74, 85.82)',
$$

and thus the fitted model is

$$
\text{price} = -1336.7 + 12.7 \times \text{age} + 85.8 \times \text{no. bidders} + \text{error}.
$$

From here, fitted values and residuals can be calculated. The covariance matrix of $\hat{\boldsymbol{\beta}}$ is

$$
\text{var}(\hat{\boldsymbol{\beta}}) = \sigma^2 \begin{bmatrix} 169545 & -773 & -5705 \\ -773 & 5 & 11 \\ -5705 & 11 & 428 \end{bmatrix} \times 10^{-5},
$$

with σ^2 being estimated by $s^2 = \sum (y_i - \hat{y}_i)^2 / (n - 2 - 1) = 17725.0$.

Of course, we would usually fit the regression using a statistical software package, the above calculations only being displayed for illustration.

10.1.2 Regression diagnostics

Once the regression equation has been fitted, model adequacy is checked. The *total sum of squares*, SS_Y, is the 'total variation' for the response variable, and is given by

$$
SS_Y = \sum_{i=1}^{n} (y_i - \bar{y})^2.
$$

This can be split into the *sum of squares due to the regression, SS_R*, and the *residual* or *error sum of squares, SS_E*. Thus

$$SS_Y = SS_R + SS_E,$$

where

$$SS_R = \sum_{j=1}^{q} \left\{ \hat{\beta}_j \sum_{i=1}^{n} (y_i - \bar{y})(x_{ij} - \bar{x}_j) \right\}, \qquad SS_E = \sum_{i=1}^{n} (y_i - \hat{y})^2.$$

It can be shown that $E(SS_E/(n - q - 1)) = \sigma^2$, and $E(SS_R/q) = \sigma^2 + \boldsymbol{\beta}'\mathbf{X}'\mathbf{X}\boldsymbol{\beta}$.
Under the hypothesis, H_0: $\boldsymbol{\beta} = \mathbf{0}$, the statistic

$$F = (SS_R/(q))/(SS_E/(n - q - 1)),$$

has an F-distribution, with $(q, n - q - 1)$ degrees of freedom. Thus an analysis of variance table can be formed to investigate whether the regression equation explains some of the variation in the responses. For the grandfather clock data, we have

Source	SS	df	MS	F	p-value
Regression	4277160	2	2138580	120.65	0.000
Error	514034	29	17725		
Total	4791194	31			

From the table we see that the model does explain much of the variation in the responses. Also, usually quoted is the 'R^2' value, which is the ratio of the regression sum of squares to the total sum of squares, and here is $4277160/4791194 = 89.3\%$, which is reasonably high in value.

The next step is to investigate whether the error term has constant variance, whether errors are uncorrelated and are normally distributed, whether there are outliers in the data, whether there are points of high-leverage, whether there are influential observations, and whether some explanatory variables can be removed from the model.

Residual plots.   The residual vector, $\mathbf{e} = \mathbf{y} - \hat{\mathbf{y}}$, has mean vector and covariance matrix given by (10.2). We can estimate the variance of the ith residual by $s\sqrt{(1 - h_{ii})}$, where h_{ii} is the ith diagonal element of $\mathbf{H}$. The *standardized* residuals are given by

$$e_i^* = \frac{e_i}{\sqrt{s^2(1 - h_{ii})}},$$

which have zero mean and variance close to unity, but are correlated with each other.

Residuals are plotted against fitted values and against each of the explanatory variables. They can also be plotted against any other relevant variable, such as time, if the observations have been collected over time. If the assumptions of the model are satisfied, then the residuals should be randomly scattered within a horizontal band. Figure 10.1 shows a plot of residuals against fitted values for the grandfather clock data. There is nothing to alarm us in the plot as to model inadequacy.

Figures 10.2 and 10.3 show the residuals plotted against the explanatory variables, x_1 and x_2, and again there is no cause for concern. Figure 10.4 shows the ordered residuals

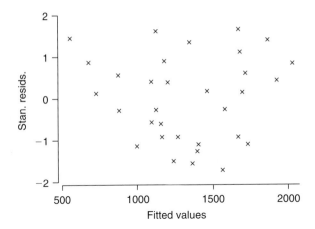

Figure 10.1 Residuals versus fitted values

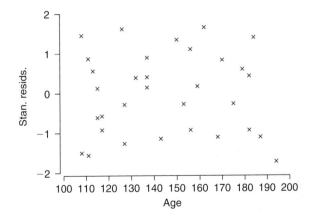

Figure 10.2 Residuals versus age

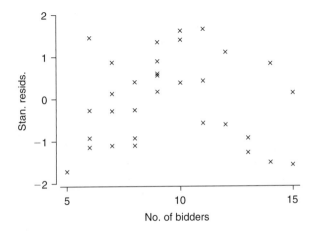

Figure 10.3 Residuals versus number of bidders

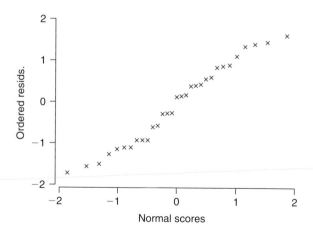

Figure 10.4 Normal plot

plotted against normal scores. Normal scores are the values of x for the standard normal distribution that give the probabilities $1/n, 2/n, \ldots, (n-1)/n$. For an adequate model the points should lie more or less on a straight line. For our data there is some concern as to normality.

Outliers. The values of the standardized residuals can indicate whether there are any discordant y-values. The observations that have large residual values should be looked at carefully, and, if necessary, dealt with by correcting errors or by deleting the observation. However, the deletion of observations should not be taken lightly, and should only be done so for good reason. The analysis can be carried out with and without the outliers, to see their effect on the analysis. How large does a standardized residual need to be in order to be classed as 'large'? A guide is ± 2 is acceptable, ± 3 is getting large, and ± 4 is cause for concern. For the grandfather clock data the largest standardized residual has the value -1.72, and hence there are no outliers.

Points of high leverage. Points of high leverage are points (observations) in the space of the explanatory variables, that have undue weight in the regression. A point, $\mathbf{x}_i$, of high leverage has a large value of h_{ii}, the ith element of the diagonal of the hat matrix, $\mathbf{H}$. Now, $n^{-1} \leq h_{ii} \leq 1$, and a value of h_{ii} greater than $2n^{-1}(q+1)$ is a guide as to whether the ith point is a point of high leverage or not. For our data, none of the points exceed the guide value, although four come close to it (points 17, 25, 27, and 31).

Influential observations. Influential observations are points which have a substantial effect on the estimate $\hat{\beta}$. To measure the effect of the ith point, the regression is carried out giving the usual estimate $\hat{\beta}$ of β. Then the ith observation is deleted, and the regression carried out again, now giving the estimate $\hat{\beta}_{(i)}$ of β. The suffix (i) is standard notation for deleting an observation. A measure of the distance between $\hat{\beta}$ and $\hat{\beta}_{(i)}$ is the Cook statistic given by

$$D_i = (q + 1)^{-1}(\hat{\boldsymbol{\beta}} - \hat{\boldsymbol{\beta}}_{(i)})' \text{vâr}(\hat{\boldsymbol{\beta}})^{-1}(\hat{\boldsymbol{\beta}} - \hat{\boldsymbol{\beta}}_{(i)})$$
$$= \{(q + 1)s^2\}^{-1}(\hat{\boldsymbol{\beta}} - \hat{\boldsymbol{\beta}}_{(i)})(\mathbf{X'X})(\hat{\boldsymbol{\beta}} - \hat{\boldsymbol{\beta}}_{(i)}).$$

After some algebra, the Cook statistic can be shown to be

$$D_i = (q + 1)^{-1} e_i^{*2} \frac{h_{ii}}{1 - h_{ii}},$$

and thus the regressions, with each observation deleted in turn, do not actually have to be carried out in practice. For our data, there are four observations that have large values of the Cook statistic compared to the others (observations 23, 25, 27 and 31).

Multicollinearity.   If there are linear dependencies among the vectors of explanatory variables, then the inverse of $(\mathbf{X'X})$ will not exist. For example, if the explanatory variables, x_1, x_2, x_3 are percentages, then $x_1 + x_2 + x_3 - 100 \; \mathbf{1} = \mathbf{0}$, and there is a linear dependence between the vectors. In practice, the vectors will probably exhibit near-linear dependency, and although $(\mathbf{X'X})^{-1}$ will exist, it will be very unstable, and the variance matrix of $\hat{\boldsymbol{\beta}}$ will contain some very large elements.

Near-linear dependencies can be detected by (i) looking at the correlations between pairs of explanatory variables, (ii) looking at the eigenvalues of $\mathbf{X'X}$, and (iii) calculating *variance inflation factors* (VIF). A very large positive or negative correlation between a pair of explanatory variables suggests a near-linear dependence between the two. A very small eigenvalue for $\mathbf{X'X}$ also suggests a near-linear dependence. A guide is that if the ratio of the largest eigenvalue to the smallest is greater than 100, then a problem of a near-dependency could exist. The coefficients of the eigenvector corresponding to the smallest eigenvalue will indicate the actual dependency. The VIF for the jth explanatory variable is $v_j = (\mathbf{X'X})^{-1} \sum_{i=1}^{n} (x_{ij} - \bar{x}_j)^2$, which measures the increase in variance of $\hat{\boldsymbol{\beta}}$, due to the near-linear dependencies. As a guide, values of VIF greater than 5 suggest near-linear dependencies.

To overcome the problem of near-linear dependence, transformations of the explanatory variables can be tried, or deletion of some of the variables, or extra observations added to the dataset.

Selection of explanatory variables.   The distribution of $\hat{\boldsymbol{\beta}}$ is $N_{q+1}(\boldsymbol{\beta}, \sigma^2(\mathbf{X'X})^{-1})$. The variance σ^2 is estimated by $s^2 = SS_E/(n - q - 1)$, where $SS_E = \sum (y_i - \hat{y}_i)^2$. Thus to test whether the jth explanatory variable is useful in the regression or not, we test the null hypothesis, $H_0: \beta_j = 0$, using the test statistic

$$t = \frac{|\hat{\beta}_j|}{\sqrt{s^2 v_{ii}}},$$

where v_{ii} is the jth diagonal element of $(\mathbf{X'X})^{-1}$. Under H_0, t has a t-distribution with $n-q-1$ degrees of freedom. Each of the explanatory variables can be tested as to their importance in the regression, but *these tests are not independent*, unless $\mathbf{X}$ is orthogonal. The tests should only be taken as a guide. For the grandfather clock data, the t-values are: constant term,

$t = 7.71$; age, $t = 14.11$, number of bidders, $t = 9.86$. These are extremely large values compared to the upper 0.01 percentile of the t_{29} distribution and so the constant and the two explanatory variables are all important in the regression.

Extra sum of squares principle.   The extra sum of squares principle can be used to test whether a subset of the explanatory variables have no effect on the response variable, *given* those variables already fitted in the regression. For convenience, the explanatory variables are arranged so that the subset that are to be tested, q_1 of them, are placed at the end, i.e. $x_{q-q_1+1}, \ldots, x_q$. Let the regression coefficients for these be denoted by $\boldsymbol{\beta}_2$, and those for the variables not under test, i.e. $x_1, \ldots, x_{q-q_1}$, by $\boldsymbol{\beta}_1$. Thus $\boldsymbol{\beta} = (\boldsymbol{\beta}_1', \boldsymbol{\beta}_2')'$.

First, fit all q explanatory variables, and from the ANOVA table, note the sum of squares due to the regression, $SS_R(\boldsymbol{\beta})$, with q df. Also note the error sum of squares, SS_E, with $(n - q - 1)$ df. Then delete the subset under test, and refit the regression, giving regression sum of squares, $SS_R(\boldsymbol{\beta}_1)$, with $(q - q_1)$ df. The extra sum of squares, by incorporating $x_{q-q_1+1}, \ldots, x_q$ into the regression, given that $x_1, \ldots, x_{q-q_1}$ are already included, is

$$SS_R(\boldsymbol{\beta}_2 | \boldsymbol{\beta}_1) = SS_R(\boldsymbol{\beta}) - SS_R(\boldsymbol{\beta}_1).$$

We test $H_0: \boldsymbol{\beta}_2 = \mathbf{0}$, using the F-statistic

$$F = \frac{SS_R(\boldsymbol{\beta}_2 | \boldsymbol{\beta}_1)/q_1}{SS_E/(n - q - 1)},$$

which has an F-distribution with $(q_1, n - q - 1)$ df, under H_0.

For the grandfather clock data, we test whether age can be withdrawn from the regression, with the number of bidders already included. We find, $SS_R(\boldsymbol{\beta}) = 4277160$, $SS_R(\boldsymbol{\beta}_1) = 746185$, and $SS_E = 514034$. Hence

$$SS_R(\boldsymbol{\beta}_2 | \boldsymbol{\beta}_1) = 4277160 - 746185 = 3530975,$$

$$F = \frac{3530975/1}{514034/29} = 199.2.$$

From the $F_{1,29}$ tables, this is seen to be a highly significant value, and hence we should retain age as an explanatory variable.

10.2 Multivariate multiple linear regression

The extension of multiple linear regression to multivariate multiple linear regression is straightforward. We now have p response variables, $Y_1, \ldots, Y_p$. The vector of responses in equation (10.1) is replaced by a matrix of responses, $\mathbf{Y}$, where the jth column contains the responses for Y_j. The vector of regression coefficients, $\boldsymbol{\beta}$, is replaced by a matrix, $\mathbf{B}$, of regression coefficients, and the error vector, by a matrix of errors, $\mathbf{E}$. The model becomes

$$\mathbf{Y} = \mathbf{XB} + \mathbf{E}.$$

The jth column of $\mathbf{B}$ contains the regression coefficients for the jth response variable, and the jth column of $\mathbf{E}$, the errors for the jth response variable.

The rth row of the error matrix contains the j errors for the rth observation, one for each of the j response variables. Denote this row of $\mathbf{E}$ by $\mathbf{e}_r$. We assume the distribution of $\mathbf{e}_r$ is $N_p(\mathbf{0}, \boldsymbol{\Sigma})$, $r = 1, \ldots, n$, and that rows of $\mathbf{E}$ are independent, so $\text{cov}(\mathbf{e}_r, \mathbf{e}_s) = \mathbf{0}$, $(r \neq s)$, where $\mathbf{0}$ is a matrix of zeros.

Similar to the univariate case, the estimates of $\mathbf{B}$ and $\boldsymbol{\Sigma}$, are given by

$$\hat{\mathbf{B}} = (\mathbf{X}'\mathbf{X})^{-1}\mathbf{X}'\mathbf{Y}$$

$$\hat{\boldsymbol{\Sigma}} = (n - q - 1)^{-1}(\mathbf{Y} - \mathbf{X}\hat{\mathbf{B}})'(\mathbf{Y} - \mathbf{X}\hat{\mathbf{B}}).$$

It can be shown that $\hat{\mathbf{B}}$ has a multivariate normal distribution, and $n\hat{\boldsymbol{\Sigma}}$ a Wishart distribution, $W_p(n - q - 1, \boldsymbol{\Sigma})$.

The matrix of fitted values, $\hat{\mathbf{Y}}$, and the matrix of residuals, $\hat{\mathbf{E}}$, are given by

$$\hat{\mathbf{Y}} = \mathbf{X}\hat{\mathbf{B}} = \mathbf{X}(\mathbf{X}'\mathbf{X})^{-1}\mathbf{X}'\mathbf{Y}, \qquad \hat{\mathbf{E}} = \mathbf{Y} - \mathbf{X}\hat{\mathbf{B}} = \mathbf{H}\mathbf{Y}.$$

An ANOVA table can be formed as for the univariate case, but this time using sums of squares and product matrices in place of SS_Y, SS_R, and SS_E,

Source	SSP	df	MSP
Regression	$\mathbf{R} = (\hat{\mathbf{Y}} - \mathbf{1}\bar{\mathbf{y}})'(\hat{\mathbf{Y}} - \mathbf{1}\bar{\mathbf{y}})$	q	$\mathbf{R}/q$
Residual	$\mathbf{W} = \hat{\mathbf{E}}'\hat{\mathbf{E}}$	$n - q - 1$	$\mathbf{W}/(n - q - 1)$
Total	$\mathbf{T} = (\mathbf{Y} - \mathbf{1}\bar{\mathbf{y}})'(\mathbf{Y} - \mathbf{1}\bar{\mathbf{y}})$	$n - 1$	

The usual F-test cannot be performed since we have matrices for the MSP column in the ANOVA table. Several alternative tests have been suggested, based on the eigenvalues, $\lambda_1, \lambda_2, \ldots, \lambda_q$ of $\mathbf{R}\mathbf{W}^{-1}$, and these are often carried out routinely by statistical software packages. The tests are:

1. Wilk's lambda, $\Lambda = \frac{|\mathbf{W}|}{|\mathbf{W} + \mathbf{R}|} = \prod_{i=1}^{q} \frac{1}{1 + \lambda_i}$
2. Roy's greatest root, λ_1
3. Lawley–Hotelling's trace, $\sum_{i=1}^{q} \lambda_i$
4. Pillai's trace, $\sum_{i=1}^{q} \frac{\lambda_i}{1 + \lambda_i}$

See Krzanowski and Marriott (1994) for a discussion on the distribution of these four statistics. Wilk's Λ will be met again in Chapter 11 in the multivariate analysis of variance. Many statistical software packages give all four of the statitics and corresponding p-values. For some special cases, appropriate transformations of Wilk's Λ have an F-distribution. However, for the general case, let r be the number of degrees of freedom for $\mathbf{R}$, w be the number of degrees of freedom for $\mathbf{W}$, and then Bartlett's approximation is that the statistic

$$-\left\{w - \tfrac{1}{2}(p - r + 1)\right\} \log \Lambda,$$

has an approximate chi-squared distribution with pr degrees of freedom. Here $r = q$ and $w = n - q - 1$.

Rao's approximation is more accurate, but is more complicated, where

$$\frac{1 - \Lambda^{1/b}}{\Lambda^{1/b}} \frac{(ab - c)}{pr}$$

has an approximate F-distribution with $(pr, ab - c)$ degrees of freedom, where

$$a = \{w - \tfrac{1}{2}(p - r + 1)\} \quad b = \{(p^2 r^2 - 4)/(p^2 + r^2 - 5)\}^{1/2} \quad c = \tfrac{1}{2}(pr - 2).$$

Note that the F-distribution can have fractional degrees of freedom.

Example: Beef and pork consumption

The DASL website has a dataset containing the beef and pork consumption in the USA, together with seven other variables, for the years 1925–1941. These classic data are useful for illustrating the technique. We will use CBE – *consumption of beef per capita* (lbs); and CPO – *consumption of pork per capita* (lbs) as the response variables and PBE – *price of beef* (cents/lb), PPO – *price of pork* (cents/lb), and DINC – *disposable income per capita index*, as the explanatory variables. The response vector and matrix of explanatory variables are

$$
\mathbf{Y} =
\begin{bmatrix}
58.6 & 65.8 \\
59.4 & 63.3 \\
53.7 & 66.8 \\
48.1 & 69.9 \\
49.0 & 68.7 \\
48.2 & 66.1 \\
47.9 & 67.4 \\
46.0 & 69.7 \\
50.8 & 68.7 \\
55.2 & 62.2 \\
52.2 & 47.7 \\
57.3 & 54.4 \\
54.4 & 55.0 \\
53.6 & 57.4 \\
53.9 & 63.9 \\
54.2 & 72.4 \\
60.0 & 67.4
\end{bmatrix}
\qquad
\mathbf{X} =
\begin{bmatrix}
1 & 59.7 & 60.5 & 51.4 \\
1 & 59.7 & 63.3 & 52.6 \\
1 & 63.0 & 59.9 & 52.1 \\
1 & 71.0 & 56.3 & 52.7 \\
1 & 71.0 & 55.0 & 55.1 \\
1 & 74.2 & 59.6 & 48.8 \\
1 & 72.1 & 57.0 & 41.5 \\
1 & 79.0 & 49.5 & 31.4 \\
1 & 73.1 & 47.3 & 29.4 \\
1 & 70.2 & 56.6 & 33.2 \\
1 & 82.2 & 73.9 & 37.0 \\
1 & 68.4 & 64.4 & 41.8 \\
1 & 73.0 & 62.2 & 44.5 \\
1 & 70.2 & 59.9 & 40.8 \\
1 & 67.8 & 51.0 & 43.5 \\
1 & 63.4 & 41.5 & 46.5 \\
1 & 56.0 & 43.9 & 56.3
\end{bmatrix}.
$$

The parameter estimates are given by

$$
\hat{\mathbf{B}} =
\begin{bmatrix}
101.4 & 79.6 \\
-0.753 & 0.153 \\
0.254 & -0.687 \\
-0.241 & 0.283
\end{bmatrix},
$$

and the matrices **R**, **W** and **T** are

$$
\mathbf{R} =
\begin{bmatrix}
235.77 & -34.24 \\
-34.24 & 487.87
\end{bmatrix}
\qquad
\mathbf{W} =
\begin{bmatrix}
57.35 & -98.43 \\
-98.43 & 218.85
\end{bmatrix}
$$

$$
\mathbf{T} =
\begin{bmatrix}
293.12 & -132.67 \\
-132.67 & 706.72
\end{bmatrix}.
$$

The eigenvalues of $\mathbf{RW}^{-1}$ are 23.77 and 1.67.

The regression equations are thus

$$\text{CBE} = 101.4 - 0.753\text{PBE} + 0.254\text{PPO} - 0.241\text{DINC}$$
$$\text{CPO} = 79.6 + 0.153\text{PBE} - 0.687\text{PPO} + 0.283\text{DINC},$$

indicating that the consumption of beef increases as the price of beef decreases and/or the price of pork increases, and with a similar relationship for the consumption of pork. The consumption of beef decreases, and the consumption of pork increases as disposable income increases.

We could have obtained these two regression equations using multiple linear regression on beef and pork consumption separately. The advantages of using multivariate multiple linear regression are that the correlation between beef consumption and pork consumption is modelled, and it also allows simultaneous confidence intervals for predicting beef and pork consumption. This could not be achieved from separate regressions.

The estimate of the error covariance matrix is

$$\hat{\Sigma} = \begin{bmatrix} 4.412 & -7.572 \\ -7.572 & 16.835 \end{bmatrix},$$

and thus the estimated error standard deviation for beef consumption is 2.10, and for pork consumption is 4.10. The estimated correlation between the errors is -0.88.

The value of Wilk's Λ is 0.414. Then for Bartlett's approximation, we have $-13 \log 0.414 = 11.46$, and from the χ_6^2 distribution, $p = 0.075$, indicating that the regression model does explain some of the variation in the data, but not as much as might have been wished for. Regression diagnostics along the lines of those described for the single response variable should be investigated, but space does not allow these to be included here.

10.3 Exercises

1. The simple linear regression model in matrix form is

$$\begin{bmatrix} y_1 \\ y_2 \\ \vdots \\ y_n \end{bmatrix} = \begin{bmatrix} 1 & x_1 \\ 1 & x_2 \\ \vdots & \vdots \\ 1 & x_n \end{bmatrix} \begin{bmatrix} \beta_0 \\ \beta_1 \end{bmatrix} + \begin{bmatrix} \varepsilon_1 \\ \varepsilon_2 \\ \vdots \\ \varepsilon_n \end{bmatrix}.$$

Use the result $\hat{\beta} = (\mathbf{X}'\mathbf{X})^{-1}\mathbf{X}'\mathbf{y}$ for multiple linear regression to obtain the usual expression for $\hat{\beta}_0$ and $\hat{\beta}_1$,

$$\hat{\beta}_1 = \frac{\sum_i (x_i - \bar{x})(y_i - \bar{y})}{\sum_i (x_i - \bar{x})^2}$$

$$\hat{\beta}_0 = \bar{y} - \hat{\beta}_1 \bar{x}.$$

2. Download the Chromotography data from the DASL website and carry out a simple linear regression analysis.

3. Hand et al. (1994) contains a dataset on the boiling point of water for various pressures in the Alps. The data can be found at the website http://www.stat.ucla.edu/. Carry out a simple linear regression analysis on the data.

4. Download the dataset on OECD Economic Development from the DASL website and carry out a multiple regression analysis.

5. Download the beef and pork consumption data from the DASL website (Agricultural Economics Studies story) and fit various models to the data. Find the correlations between the two consumption variables and the regressor variables in order to make a choice of which regressors to use.

   Look at some of the regression diagnostics.

   Fit models to beef consumption and pork consumption separately and compare the results with those for the combined model.

11

Multivariate analysis of variance

In this chapter we look at the generalization of the analysis of variance (ANOVA) to the multivariate case (MANOVA). It is assumed the reader is familiar with the basic ideas of the design and analysis of experiments, and in particular, the one-factor model. First, we give an example of the univariate one-factor model and then extend it to the multivariate case. Space does not allow extension to two or more factor models, or other designs.

11.1 Univariate one-factor model

After the Roman invasion of Britain, the British pottery industry expanded and improved under the influence of the Romans, and hence the term *Romano-British* pottery. The DASL website contains an archaelogical dataset relating to the percentage of five oxides contained in Romano-British pottery found at four different sites. The oxides are aluminium, iron, magnesium, calcium and sodium. The sites where the pottery was found are Llanederyn, Caldicot, Island Thorns and Ashley Rails. Llanederyn and Caldicot are near Newport in Wales, and Ashley Rails and Island Thorns are in Hampshire. This dataset is widely used to illustrate MANOVA. The means and standard deviations and sample sizes of the oxides at the four sites are given in Table 11.1.

First, we review standard ANOVA by considering the aluminium content only. The one-factor model for k treatment levels is

$$y_{ij} = \mu + \tau_i + \varepsilon_{ij} \quad (i = 1, \ldots, k; \ j = 1, \ldots, n_i),$$

where μ is the overall mean, τ_i is the effect due to the ith level (site), and ε_{ij} is the 'error', assumed to be normally distributed with mean zero and variance σ^2. Also, observations are assumed to be independent. The model is over-parameterized, and so we impose the restriction $\sum n_i \tau_i = 0$, or we can make one of the τ_i's equal to zero. We choose the latter ($\tau_k = 0$).

The *fundamental equation of analysis of variance* is

$$SS_Y = SS_T + SS_E,$$

Table 11.1 Means, standard deviations and sample size for the pottery data

Site	N	Al mean	Al sd	Fe mean	Fe sd	Mg mean	Mg sd
Llanederyn	14	12.56	1.38	6.37	0.79	4.83	1.09
Caldicot	2	11.70	0.14	5.42	0.04	3.86	0.12
Island Thorns	5	18.18	1.78	1.71	0.44	0.67	0.03
Ashley Rails	5	17.32	1.66	1.51	0.74	0.61	0.06

Site	N	Ca mean	Ca sd	Na mean	Na sd
Llanederyn	14	0.20	0.06	0.25	0.12
Caldicot	2	0.30	0.01	0.05	0.01
Island Thorns	5	0.03	0.03	0.05	0.03
Ashley Rails	5	0.05	0.03	0.05	0.01

where SS_Y is the *total (corrected) sum of squares*, SS_T is the *treatment sum of squares*, and SS_E is the *error sum of squares*. These three quantities are calculated as

$$SS_Y = \sum_{i=1}^{k} \sum_{j=1}^{n_i} (y_{ij} - \bar{y}_{..})^2$$

$$SS_T = \sum_{i=1}^{k} \sum_{j=1}^{n_i} (\bar{y}_{i.} - \bar{y}_{..})^2$$

$$SS_E = \sum_{i=1}^{k} \sum_{j=1}^{n_i} (y_{ij} - \bar{y}_{i.})^2,$$

where the dot suffix, together with the bar, means to take the average over that suffix with the dot. Let $N = \sum_{i=1}^{k} n_i$.

The *error mean square* is given by $MS_E = SS_E/(N-k)$, and is an unbiased estimator of the population variance σ^2. The *treatment mean square* is given by $MS_T = SS_T/(k-1)$. Under the null hypothesis that all the treatment means are equal,

$$H_0 : \tau_1 = \tau_2 = \ldots = \tau_k = 0,$$

MS_T is also an unbiased estimator of σ^2. The *F ratio* is used to test H_0, where

$$F = \frac{MS_T}{MS_E}.$$

Under the null hypothesis, F has an F-distribution with degrees of freedom $v_1 = k - 1$, and $v_2 = N - k$. An ANOVA table is formed as follows and H_0 tested.

Source	SS	df	MS	F	p-value
Treatment	SS_T	$k-1$	MS_T	$f = MS_T/MS_E$	$P(F_{k-1, N-k} \geq f)$
Error	SS_E	$N-k$	MS_E		
Total	SS_Y	$N-1$			

The ANOVA table for the aluminium content for the Romano-British pottery is

Source	SS	df	MS	F	p-value
Treatment	175.61	3	58.54	26.67	<0.0001
Error	48.29	22	2.19		
Total	223.90	25			

giving overwhelming evidence of differences in aluminium content at the various sites.

The estimated mean aluminium oxide content at the sites are: Llanederyn 12.56, Caldicot 11.70, Island Thorns 18.18, and Ashley Rails 17.32. *Multiple comparisons* attempt to find where the differences lie, between the means for the k levels. There are several different methods for multiple comparisons, and there is much discussion as to which methods are the most appropriate (see Hsu, 1996). Here, we use Student's t-tests between all six pairs of population means, but using a common estimate of σ for the denominator in the t-statistic. Note that the tests are conditioned on the fact that we have already rejected the null hypothesis, H_0 above, of equal population means. If we were to carry out these six t-tests without first testing H_0, then we would increase the overall significance level of our test substantially. Essentially, if we carry out many hypothesis tests, we will surely find some differences between means by chance, even when there is no difference between the population means. To overcome this, a Bonferroni adjustment can be made where the significance level, α, used for testing each pair of means, is replaced by the value α/k. For our data, multiple comparison tests showed no significant difference between the population means for Ashley Rails (A) and Island Thorns (I), nor between those for Llanederyn (L) and Caldicot (C), but significant differences between A and L, A and C, I and L, and I and C. Thus A and I are a closely related pair, and so are L and C.

Particular hypotheses about the means can be tested using *linear contrasts*, which test a linear combination of the population means. For example, given two of the archaeological sites are in Wales and two are in Hampshire, we might wish to test whether the population means for aluminium oxide are different for the two areas. Assuming the four sites are put in alphabetical order for analysis, Ashley Rails, Caldicot, Island Thorns, and Llanederyn, then the linear contrast to be tested is $\mathbf{c} = (1, -1, 1, -1)'$, and placing the four population means into a vector, $\boldsymbol{\mu}$, the hypothesis to be tested is

$$H_0 : \mathbf{c}'\boldsymbol{\mu} = 0.$$

The test statistic is $\bar{y}_1 - \bar{y}_2 + \bar{y}_3 - \bar{y}_4 = \mathbf{c}'\bar{\mathbf{y}}$ which has estimated standard error, $\hat{\sigma}\sqrt{(n_1^{-1} + n_2^{-1} + n_3^{-1} + n_4^{-1})}$. For the aluminium oxide content, $\mathbf{c}'\bar{\mathbf{y}} = 11.236$ and the estimated standard error is 0.971. The hypothesis of equal aluminium oxide content for the two areas is thus emphatically rejected.

The general linear contrast is $c = (c_1, c_2, \ldots, c_k)'$, with $\sum_{i=1}^{k} c_i = 0$. Several contrasts can be specified to be tested, ideally before any data are collected. Results for several contrasts will not be independent, unless they are *orthogonal*. Two contrasts, c and d, are orthogonal if $\sum_{i=1}^{k} c_i d_i / n_i = 0$.

11.2 Multivariate one-factor model

We now generalize the univariate one-factor model to a multivariate model. The advantage of the multivariate model is that it allows the testing of several variables simultaneously, and thus avoiding multiplicity problems in testing each variable individually. For the multivariate model, the response variable y in the univariate model is simply replaced by a vector response, $\mathbf{y} = (y_1, \ldots, y_p)$, the parameters μ and τ are replaced by parameter vectors, and the error ε is replaced by an error vector. Thus the model is

$$\mathbf{y}_{ij} = \boldsymbol{\mu} + \boldsymbol{\tau}_i + \boldsymbol{\varepsilon}_{ij} \quad (i = 1, \ldots, k; \; j = 1, \ldots, n_i).$$

The error vector is assumed to have a multivariate normal distribution, $\boldsymbol{\varepsilon}_{ij} \sim N_p(\mathbf{0}, \boldsymbol{\Sigma})$, and again, observations are independent.

For each element y_l of $\mathbf{y}$, we can find the total corrected sum of squares as for the univariate case. We can also find, for each pair of elements, $y_l, y_{l'}$, the total sum of corrected cross-products, $\sum_i \sum_j (y_{(l)ij} - \bar{y}_{(l)})(y_{(l')ij} - \bar{y}_{(l')})$, noting the use of (l) and (l') to denote the lth and l'th elements of $\mathbf{y}$. All of the total sums of squares and cross-products are placed in a matrix $\mathbf{T}$,

$$\mathbf{T} = \sum_{i=1}^{k} \sum_{j=1}^{n_i} (\mathbf{y}_{ij} - \bar{\mathbf{y}})(\mathbf{y}_{ij} - \bar{\mathbf{y}})',$$

where $\bar{\mathbf{y}} = \sum_{i=1}^{k} \sum_{j=1}^{n_i} \mathbf{y}_{ij} / N$ is the overall mean vector. In general, we will use 'SSP' to refer to a sums of squares and cross-products matrix.

The treatment SSP is given by

$$\mathbf{B} = \sum_{i=1}^{k} n_i (\bar{\mathbf{y}}_i - \bar{\mathbf{y}})(\bar{\mathbf{y}}_i - \bar{\mathbf{y}})',$$

where $\bar{\mathbf{y}}_i = \sum_{j=1}^{n_i} \mathbf{y}_{ij} / n_i$. Note we have used the letter '$\mathbf{B}$' for the treatment SSP, as sometimes it is referred to as the 'between' treatments SSP.

The error SSP is given by

$$\mathbf{W} = \sum_{i=1}^{k} \sum_{j=1}^{n_i} (\mathbf{y}_{ij} - \bar{\mathbf{y}}_i)(\mathbf{y}_{ij} - \bar{\mathbf{y}}_i)',$$

this time using the letter '$\mathbf{W}$' as it is often referred to as the 'within' treatments SSP.

As for the univariate case, we have the fundamental equation

$$\mathbf{T} = \mathbf{W} + \mathbf{B}.$$

An ANOVA table is formed as before, but with matrices rather than scalars:

Source	SSP	df	Wilks' λ
Treatment	**B**	$k-1$	$|\mathbf{W}|/|\mathbf{W}+\mathbf{B}|$
Error	**W**	$N-k$	
Total	**T**	$N-1$	

The hypothesis usually tested is

$$H_0: \tau_1 = \tau_2 = \ldots = \tau_k,$$

against the alternative that at least one pair $(\tau_i, \tau_{i'})$ are unequal. We can no longer use the F-test, since we cannot simply divide one matrix by another, but instead, we use Wilks' Λ, with $(p, n-k, k-1)$ degrees of freedom (see Chapter 10). This is the test statistic obtained from the generalized likelihood ratio test when testing $H_0: \mu_1 = \mu_2 = \ldots = \mu_k$, against $H_1: \mu$'s are different; see Mardia et al. (1979) for further details.

11.2.1 Romano-British pottery continued

We are now in a position to carry out a MANOVA for the Romano-British pottery data. The total SSP matrix is

$$\mathbf{T} = \begin{pmatrix} 223.90 & -142.22 & -130.20 & -5.78 & -4.78 \\ -142.22 & 145.17 & 118.27 & 4.67 & 5.39 \\ -130.20 & 118.27 & 118.78 & 4.64 & 4.74 \\ -5.78 & 4.67 & 4.64 & 0.26 & 0.16 \\ -4.78 & 5.39 & 4.74 & 0.16 & 0.46 \end{pmatrix}.$$

The treatment SSP is

$$\mathbf{B} = \begin{pmatrix} 175.61 & -149.30 & -130.81 & -5.89 & -5.37 \\ -149.30 & 134.22 & 117.75 & 4.82 & 5.33 \\ -130.81 & 117.75 & 103.35 & 4.21 & 4.71 \\ -5.89 & 4.82 & 4.21 & 0.20 & 0.15 \\ -5.37 & 5.33 & 4.71 & 0.15 & 0.26 \end{pmatrix}.$$

The error SSP is

$$\mathbf{B} = \begin{pmatrix} 48.29 & 7.08 & 0.61 & 0.11 & 0.59 \\ 7.08 & 10.95 & 0.53 & -0.16 & 0.07 \\ 0.61 & 0.53 & 15.43 & 0.44 & 0.03 \\ 0.11 & -0.16 & 0.44 & 0.05 & 0.01 \\ 0.59 & 0.07 & 0.03 & 0.01 & 0.20 \end{pmatrix}.$$

Wilks' Λ can be calculated as

$$\Lambda = \prod_{i=1}^{p} (1 + \lambda_i)^{-1},$$

where $\lambda_1, \ldots, \lambda_p$ are the eigenvalues of $\mathbf{W}^{-1}\mathbf{B}$. For the above matrices, these eigenvalues are 34.16, 1.25, 0.03, 0.0, and 0.0. Thus $\Lambda = 0.0123$, and then Rao's approximation gives $F = 13.09$. The degrees of freedom for the F-distribution are (15, 50.1), and hence $p < 0.0001$, indicating that there are differences between the mineral contents at the various sites.

11.2.2 Multivariate Bartlett test

One assumption in MANOVA is that the groups have a common covariance matrix, $\mathbf{\Sigma}$. Bartlett's test for testing homogeneity of variance for the univariate case can be extended to the multivariate case. The test statistic to test the hypothesis

$$H_0: \mathbf{\Sigma}_1 = \mathbf{\Sigma}_2 = \ldots = \mathbf{\Sigma}_k,$$

is

$$\log M = \frac{1}{2} \sum_{i=1}^{k} (n_i - 1) \log |\mathbf{S}_i| - \frac{1}{2} \left(\sum_{i=1}^{k} (n_i - 1) \right) \log |\mathbf{S}_u|,$$

where $\mathbf{S}_i$ is the sample covariance matrix for the ith group, and $\mathbf{S}_u$ is the pooled sample covariance matrix, $\mathbf{S}_u = \sum_i (n_i - 1)\mathbf{S}_i / (\sum_i n_i - k)$. Let $u = -2(1 - c) \log M$, where

$$c = \left[\sum_{i=1}^{k} (n_i - 1)^{-1} - \left(\sum_{i=1}^{k} (n_i - 1) \right)^{-1} \right] \left[\frac{2p^2 + 3p - 1}{6(p + 1)(k - 1)} \right].$$

Then under H_0, and multivariate normality, u has an approximate chi-squared distribution, with $\frac{1}{2}(k - 1)p(p + 1)$ degrees of freedom.

11.3 Profile analysis

Figure 11.1 shows a profile plot for the four archaeological sites. The profile for each site is a plot of the mean values for the five oxides, connected by straight lines. The order of the plotting of the five oxides will affect the appearance of the profile, but in some situations there will be a natural ordering, for instance, if there were a time sequence to the variables. It appears that the profiles of Ashley Rails and Island Thorns are similar to each other, as are those of Caldicot and Llanederyn.

We are interested in whether there are any differences between the profiles. Firstly, are the profiles *parallel*? Note that this only makes sense when all the variables are measured on the same scale. Let the k mean vectors be placed in a matrix $\mathbf{\Delta}$,

$$\mathbf{\Delta} = (\boldsymbol{\mu}_1, \ldots, \boldsymbol{\mu}_k)' = \begin{pmatrix} \mu_{11} & \mu_{12} & \cdots & \mu_{1p} \\ \mu_{21} & \mu_{22} & \cdots & \mu_{2p} \\ \vdots & \vdots & \ddots & \vdots \\ \mu_{k1} & \mu_{k2} & \cdots & \mu_{kp} \end{pmatrix}.$$

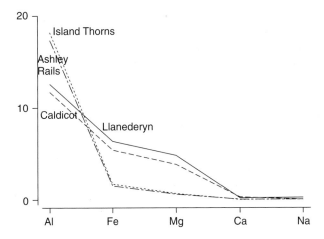

Figure 11.1 Profile plot of the sites

The first and second profiles are parallel, if the distance between μ_{11} and μ_{21} is maintained along the whole length of the profile, and thus we would have

$$\mu_{11} - \mu_{21} = \mu_{12} - \mu_{22} = \mu_{13} - \mu_{23} = \ldots = \mu_{1p} - \mu_{2p}.$$

For all the profiles to be parallel, this relationship should hold for profiles 2 and 3, profiles 3 and 4, etc. Hence the hypothesis for parallel profiles can be written as

$$H_{\text{parallel}} : \begin{pmatrix} \mu_{11} - \mu_{21} \\ \mu_{21} - \mu_{31} \\ \vdots \\ \mu_{k-1,1} - \mu_{k,1} \end{pmatrix} = \ldots = \begin{pmatrix} \mu_{1p} - \mu_{2p} \\ \mu_{2p} - \mu_{3p} \\ \vdots \\ \mu_{k-1,p} - \mu_{k,p} \end{pmatrix}.$$

Let the matrices $\mathbf{C}$ $((k-1) \times k)$ and $\mathbf{M}$ $(p \times (p+1))$, be

$$\mathbf{C} = \begin{bmatrix} 1 & -1 & 0 & \ldots & 0 \\ 0 & 1 & -1 & \ldots & 0 \\ \vdots & \vdots & \vdots & & \vdots \\ 0 & 0 & 0 & \ldots & -1 \end{bmatrix} \quad \mathbf{M} = \begin{bmatrix} 1 & 0 & \ldots & 0 \\ -1 & 1 & \ldots & 0 \\ \vdots & \vdots & & \vdots \\ 0 & 0 & \ldots & -1 \end{bmatrix}.$$

Then the parallel profiles hypothesis can be written as

$$H_{\text{parallel}} : \mathbf{C\Delta M} = \mathbf{0}.$$

To test H_0, we can postmultiply the response matrix, $\mathbf{Y}$, by $\mathbf{M}$, i.e. take successive differences in the response vectors, and carry out a one-way multivariate analysis of variance on these differences. In practice, to carry out the analysis using a statistical software package, for example SAS, the matrix $\mathbf{M}$ is specified as a transformation to be applied to the responses, and $\mathbf{C}$ as a series of linear contrasts. Using SAS, the value of Wilk's lambda is 0.019 ($F = 14.57$, df $= 12$, 50.6, $p < 0.0001$) and hence we reject the hypothesis of parallel profiles. We can carry out a similar test, to test whether there are two pairs of parallel profiles, one pair for Ashley Rails and Island Thorns, and one pair for Caldicot and Llanederyn, by choosing **C** appropriately, and the corresponding linear contrasts for SAS. (Note, the hypothesis of different profiles for Welsh and English sites was set a priori. The analyst must be wary against data dredging, choosing hypotheses to be tested based on analyses carried out on the data.) The value of Wilk's lambda for testing two pairs of profiles is 0.477 ($F = 2.13$, df $= 8$, 38, $p = 0.057$), and hence the decision is marginal as to whether we reject the hypothesis or not.

In general, if the hypothesis of parallel profiles is accepted, we are then interested in whether the profiles are *coincident*. We can test for coincident profiles by testing whether the sum of the mean values of the p variables for each treatment are equal, i.e.

$$H_{coincident} : \sum_{k=1}^{p} \mu_{1k} = \ldots = \sum_{k=1}^{p} \mu_{hk}.$$

The hypothesis can also be expressed as $H_{coincident}$: $C\mathbf{A}1 = 0$. This hypothesis is tested using a one-way analysis of variance on the sum of the responses for each observation, i.e. on the row sums of the data matrix. For the Romano-British pottery, this test is not appropriate, because we did not accept the hypothesis of parallel profiles, although we could possibly do it for the pairs of profiles.

Lastly, in general, if the hypothesis of coincident profiles is accepted, then we can test whether the means for the p response variables are equal, implying *level profiles*. The hypothesis is

$$H_{level} : \sum_{i=1}^{k} \mu_{i1} = \ldots = \sum_{i=1}^{k} \mu_{ip}.$$

The hypothesis can also be expressed as H_{level}: $1'\mathbf{A}\mathbf{M} = 0$. As before, appropriate transformations and contrasts can be set to test this hypothesis.

11.4 Exercises

1. Download the Romano-British pottery dataset from the DASL website.
   Carry out an analysis of variance for each of the oxides.
   Use the multivariate Bartlett test for homogeneity of the covariance matrices.
   Carry out a multivariate analysis of variance on the data.
   Show that the following contrasts are orthogonal, and estimate them.

Contrast	Ashley Rails	Caldicot	Island Thorns	Llanederyn
c_1	$\frac{8}{16}$	$-\frac{2}{16}$	$\frac{8}{16}$	$-\frac{14}{16}$
c_2	1	0	-1	0
c_3	0	1	0	-1

2. The following table gives the systolic and diastolic blood pressure for fifty-six 50-year-old patients. The patients are classified by gender and whether they smoke or not (s, ns).

Female	s	Sys	150	160	236	120	160	166	172
		Dia	106	90	140	82	90	110	110
		Sys	134	134	290	160	110	135	110
		Dia	80	86	124	100	75	85	70
Female	ns	Sys	150	118	180	160	195	220	180
		Dia	96	80	105	90	125	110	100
		Sys	128	168	198	140	175	156	188
		Dia	90	96	120	84	88	82	128
Male	s	Sys	160	130	118	135	108	124	120
		Dia	108	98	70	98	82	84	76
		Sys	165	140	130	130	148	120	142
		Dia	95	88	90	80	95	74	80
Male	ns	Sys	170	220	190	140	134	120	136
		Dia	104	120	104	84	88	86	86
		Sys	110	140	132	126	164	145	124
		Dia	70	86	86	90	104	120	80

The two response variables are systolic and diastolic pressure. Find mean vectors and covariance matrices for each of the four groups,

female smoker,   female non-smoker,   male smoker,   male non-smoker.

Carry out a MANOVA for the four groups (Female, s; Female, ns; Male, s; Male, ns). Note that these data could be analysed using a two-factor model.

12

Canonical correlation analysis

Suppose we have multivariate data, such that the data matrix contains variables that fall into two distinct groups. For instance, data collected on patients may split into that for physiological variables, and that for psychological variables. The results for the decathlon event at an athletics meeting fall into the results for track events (100 metres race, 400 metres race, 110 metres hurdles race, and 1500 metres race), and those for the field events (long jump, shot put, high jump, discus, pole vault, and javelin). Canonical correlation analysis attempts to measure the relationship between these two sets of variables.

There are several ways in which the relationship between the two sets of variables can be measured. The simplest is to calculate the sample correlation matrix between all the variables, and attempt an interpretation. For example, the sample correlation matrix for the results for the decathlon, in the World Athletics Championships held at Edmonton (http://www.iaaf.org/), is

	100 m	LJ	SP	HJ	400 m	110 h	DIS	PV	JAV	1500 m
100 m	1.0	.	.	.	.	.	.	.	.	.
LJ	0.70	1.0	.	.	.	.	.	.	.	.
SP	0.38	0.46	1.0	.	.	.	.	.	.	.
HJ	0.17	0.18	0.23	1.0	.	.	.	.	.	.
400 m	0.67	0.55	0.06	0.20	1.0	.	.	.	.	.
110 h	0.30	0.08	−0.01	−0.03	0.10	1.0	.	.	.	.
DIS	−0.52	−0.04	0.29	0.38	−0.38	−0.38	1.0	.	.	.
PV	0.11	−0.18	−0.13	−0.13	0.05	0.34	−0.35	1.0	.	.
JAV	0.27	0.52	0.10	0.03	0.18	−0.21	0.04	0.11	1.0	.
1500 m	−0.46	−0.42	−0.34	0.20	0.06	−0.30	0.30	0.22	−0.10	1.0

The results for the track events are measured in seconds, and the results for the field events are measured in metres. Since each athlete tries to minimize time for track events, but maximize distance for field events, in order to put all measurements on an equal footing, the distances for field events have been negated. Thus, if athletes who are good at the 100 m are also good at the long jump, then the correlation of the scores for the 100 m and the long jump will be positive, rather than negative. Indeed, the correlation for the 100 m and the long jump is high at 0.70, which is not surprising, given the nature of the two events.

Looking for relationships between the results for the track and those for the field events, it can be seen that the results for the 100 m are positively correlated with those for the long jump, but negatively correlated with those for the discus. This is also true for the 400 m event. The results for the 110 m hurdles are negatively correlated with those for the discus, but positively correlated with those for the pole vault. The 1500 m race is the last event, and often athletes will more or less know their final positions at this stage, which can lead to demotivation. This in turn can lead to some spurious results and correlations between the results for this event and the others. For example, the correlation of the results for the 1500 m with those for the discus is 0.30, and although not a particularly large correlation, is surprising nonetheless. The results for the 1500 m are also negatively correlated with those for the long jump and shot put.

If the number of variables was large, an overall interpretation of the relationship between the two groups of variables would be difficult. Canonical correlation analysis simplifies the interpretation, by finding the linear combination of the variables in the first group of variables, and the linear combination of the variables in the second group, which together have the largest possible correlation. Then other such linear combinations are found that also have maximum correlation, but with the constraint of being orthogonal to the linear combinations already found.

12.1 Mathematical development

Let $X_1, X_2, \ldots, X_p$ be the variables in the first group (e.g. the track event times), and placed in random vector $\mathbf{x}$. Let $Y_1, Y_2, \ldots, Y_q$ be the variables in the second group of variables (e.g. the negated field event distances), and placed in random vector $\mathbf{y}$. Let $U = \mathbf{a}'\mathbf{x}$ be an arbitrary linear combination of the X variables, and $V = \mathbf{b}'\mathbf{y}$ be an arbitrary linear combination of the Y variables. Then, $\mathbf{a}$ and $\mathbf{b}$ are found such that U and V have maximum correlation.

Let the covariance matrix of $(\mathbf{x}, \mathbf{y})$ be

$$\Sigma = \begin{pmatrix} \Sigma_{11} & \Sigma_{12} \\ \Sigma_{21} & \Sigma_{22} \end{pmatrix},$$

(with $\Sigma'_{21} = \Sigma_{12}$). Then the standard deviation of U is $(\mathbf{a}'\Sigma_{11}\mathbf{a})^{1/2}$, and that of V is $(\mathbf{b}'\Sigma_{22}\mathbf{b})^{1/2}$. The covariance of U and V is $\mathbf{a}'\Sigma_{12}\mathbf{b}$, and thus the correlation, $\rho(\mathbf{a}, \mathbf{b})$, between U and V is

$$\rho(\mathbf{a}, \mathbf{b}) = \frac{\mathbf{a}'\Sigma_{12}\mathbf{b}}{\{(\mathbf{a}'\Sigma_{11}\mathbf{a})(\mathbf{b}'\Sigma_{22}\mathbf{b})\}^{1/2}}.$$

We now need to find $\mathbf{a}$ and $\mathbf{b}$ such that $\rho(\mathbf{a}, \mathbf{b})$ is maximized.

Now $\rho(\mathbf{a}, \mathbf{b})$ does not depend on the scale of $\mathbf{a}$ or $\mathbf{b}$, and hence we can rescale $\mathbf{a}$ and $\mathbf{b}$ for our convenience. We choose $\mathbf{a}$ and $\mathbf{b}$, so that

$$\mathbf{a}'\Sigma_{11}\mathbf{a} = \mathbf{b}'\Sigma_{22}\mathbf{b} = 1.$$

So, the problem now is to maximize $\rho(\mathbf{a}, \mathbf{b})$, subject to $\mathbf{a}'\Sigma_{11}\mathbf{a} = \mathbf{b}'\Sigma_{22}\mathbf{b} = 1$. This is done using Lagrange multipliers. Let

$$\rho = \mathbf{a}'\Sigma_{12}\mathbf{b} - \tfrac{1}{2}\lambda(\mathbf{a}'\Sigma_{11}\mathbf{a} - 1) - \tfrac{1}{2}\gamma(\mathbf{b}'\Sigma_{22}\mathbf{b} - 1).$$

Then, differentiating, and equating to $\mathbf{0}$,

$$\frac{\partial \rho}{\partial \mathbf{a}} = \mathbf{\Sigma}_{12}\mathbf{b} - \lambda \mathbf{\Sigma}_{11}\mathbf{a} = \mathbf{0} \tag{12.1}$$

$$\frac{\partial \rho}{\partial \mathbf{b}} = \mathbf{\Sigma}_{21}\mathbf{a} - \gamma \mathbf{\Sigma}_{22}\mathbf{b} = \mathbf{0}. \tag{12.2}$$

Premultiply equation (12.1) by $\mathbf{a}'$, and equation (12.2) by $\mathbf{b}'$ to give

$$\mathbf{a}'\mathbf{\Sigma}_{12}\mathbf{b} - \lambda \mathbf{a}'\mathbf{\Sigma}_{11}\mathbf{a} = \mathbf{0}$$

$$\mathbf{b}'\mathbf{\Sigma}_{21}\mathbf{a} - \gamma \mathbf{b}'\mathbf{\Sigma}_{22}\mathbf{b} = \mathbf{0}.$$

Thus $\lambda = \gamma = \mathbf{a}'\mathbf{\Sigma}_{12}\mathbf{b} = \rho$, and equations (12.1) and (12.2) become

$$\mathbf{\Sigma}_{12}\mathbf{b} - \rho \mathbf{\Sigma}_{11}\mathbf{a} = \mathbf{0} \tag{12.3}$$

$$\mathbf{\Sigma}_{21}\mathbf{a} - \rho \mathbf{\Sigma}_{22}\mathbf{b} = \mathbf{0}. \tag{12.4}$$

Now premultiply equation (12.3) by $\rho \mathbf{\Sigma}_{11}^{-1}$, and equation (12.4) by $\mathbf{\Sigma}_{11}^{-1}\mathbf{\Sigma}_{12}\mathbf{\Sigma}_{22}^{-1}$ to obtain

$$\rho \mathbf{\Sigma}_{11}^{-1}\mathbf{\Sigma}_{12}\mathbf{b} - \rho^2 \mathbf{I}\mathbf{a} = \mathbf{0} \tag{12.5}$$

$$\mathbf{\Sigma}_{11}^{-1}\mathbf{\Sigma}_{12}\mathbf{\Sigma}_{22}^{-1}\mathbf{\Sigma}_{21}\mathbf{a} - \rho \mathbf{\Sigma}_{11}^{-1}\mathbf{\Sigma}_{12}\mathbf{b} = \mathbf{0}. \tag{12.6}$$

Adding equations (12.5) and (12.6), gives

$$(\mathbf{\Sigma}_{11}^{-1}\mathbf{\Sigma}_{12}\mathbf{\Sigma}_{22}^{-1}\mathbf{\Sigma}_{21} - \rho^2 \mathbf{I})\mathbf{a} = \mathbf{0},$$

and thus ρ^2 is an eigenvalue of $\mathbf{\Sigma}_{11}^{-1}\mathbf{\Sigma}_{12}\mathbf{\Sigma}_{22}^{-1}\mathbf{\Sigma}_{21}$, with $\mathbf{a}$ the corresponding eigenvector. In like manner,

$$(\mathbf{\Sigma}_{22}^{-1}\mathbf{\Sigma}_{21}\mathbf{\Sigma}_{11}^{-1}\mathbf{\Sigma}_{12} - \rho^2 \mathbf{I})\mathbf{b} = \mathbf{0},$$

and so ρ^2 is also an eigenvalue of $\mathbf{\Sigma}_{22}^{-1}\mathbf{\Sigma}_{21}\mathbf{\Sigma}_{11}^{-1}\mathbf{\Sigma}_{12}$, with $\mathbf{b}$ the corresponding eigenvector.

We need to look at the nature of these eigenvalues. Write the matrix $\mathbf{\Sigma}_{11}^{-1}\mathbf{\Sigma}_{12}\mathbf{\Sigma}_{22}^{-1}\mathbf{\Sigma}_{21}$, as

$$\mathbf{\Sigma}_{11}^{-\frac{1}{2}}(\mathbf{\Sigma}_{11}^{-\frac{1}{2}}\mathbf{\Sigma}_{12}\mathbf{\Sigma}_{22}^{-\frac{1}{2}})(\mathbf{\Sigma}_{11}^{-\frac{1}{2}}\mathbf{\Sigma}_{12}\mathbf{\Sigma}_{22}^{-\frac{1}{2}})'\mathbf{\Sigma}_{11}^{\frac{1}{2}},$$

which is of the form $\mathbf{\Sigma}_{11}^{-\frac{1}{2}}\mathbf{C}\mathbf{C}'\mathbf{\Sigma}_{11}^{\frac{1}{2}}$. Recalling that the non-zero eigenvalues of $\mathbf{AB}$ are the same as those for $\mathbf{BA}$, the non-zero eigenvalues of

$$\mathbf{\Sigma}_{11}^{-\frac{1}{2}}(\mathbf{\Sigma}_{11}^{-\frac{1}{2}}\mathbf{\Sigma}_{12}\mathbf{\Sigma}_{22}^{-\frac{1}{2}})(\mathbf{\Sigma}_{11}^{-\frac{1}{2}}\mathbf{\Sigma}_{12}\mathbf{\Sigma}_{22}^{-\frac{1}{2}})'\mathbf{\Sigma}_{11}^{\frac{1}{2}},$$

are the same as those for

$$(\mathbf{\Sigma}_{11}^{-\frac{1}{2}}\mathbf{\Sigma}_{12}\mathbf{\Sigma}_{22}^{-\frac{1}{2}})(\mathbf{\Sigma}_{11}^{-\frac{1}{2}}\mathbf{\Sigma}_{12}\mathbf{\Sigma}_{22}^{-\frac{1}{2}})'.$$

But, as this matrix is of the form $\mathbf{C}\mathbf{C}'$, it is positive semi-definite, and all its eigenvalues are non-negative. Thus, the maximum correlation, ρ, is taken as the square root of the largest eigenvalue of $\mathbf{\Sigma}_{11}^{-1}\mathbf{\Sigma}_{12}\mathbf{\Sigma}_{22}^{-1}\mathbf{\Sigma}_{21}$, or equivalently of $\mathbf{\Sigma}_{22}^{-1}\mathbf{\Sigma}_{21}\mathbf{\Sigma}_{11}^{-1}\mathbf{\Sigma}_{12}$, again using the fact that the non-zero eigenvalues of $\mathbf{AB}$ are the same as those for $\mathbf{BA}$. The positive or negative square root can be used, since the correlation of U and V will be positive or negative according to the choice of eigenvectors $\mathbf{a}$ and $\mathbf{b}$, as either $\mathbf{a}$ or $-\mathbf{a}$, and similarly for $\mathbf{b}$.

Let this largest correlation be relabelled ρ_1, with corresponding linear combinations, $U_1 = \mathbf{a}_1'\mathbf{x}$, $V_1 = \mathbf{b}_1'\mathbf{y}$. Then, ρ_1 is called the first canonical correlation coefficient, and U_1 and V_1 are the first canonical correlation variables. We can go on to find two more linear combinations, $U_2 = \mathbf{a}_2'\mathbf{x}$, and $V_2 = \mathbf{b}_2'\mathbf{y}$, which have maximum correlation with each other, but are also uncorrelated with the first canonical correlation variables. It can be shown that this second maximum correlation is given by the square root of the second largest eigenvalue of $\Sigma_{11}^{-1}\Sigma_{12}\Sigma_{22}^{-1}\Sigma_{21}$, or $\Sigma_{22}^{-1}\Sigma_{21}\Sigma_{11}^{-1}\Sigma_{12}$ with $\mathbf{a}_2$, and $\mathbf{b}_2$ being the corresponding eigenvectors. These are the second canonical correlation variables, with the second canonical correlation coefficient. This process can be continued until a set of $\min(p,q)$ is found.

These canonical correlation variables are population canonical correlation variables, since the population covariance matrix has been used in the derivation. In practice, this will not usually be known, and so sample canonical correlation variables are found, by replacing the population covariance matrix by the sample covariance matrix. Thus eigenvalues and eigenvectors of $\mathbf{S}_{11}^{-1}\mathbf{S}_{12}\mathbf{S}_{22}^{-1}\mathbf{S}_{21}$ and $\mathbf{S}_{22}^{-1}\mathbf{S}_{21}\mathbf{S}_{11}^{-1}\mathbf{S}_{12}$ are found.

It is possible to test whether the population correlation canonical correlations are zero by setting up the following null hypothesis:

H_0: the tth canonical correlation and all that follow $= 0$.

This is tested for $t = 1, 2, \ldots, r$, until non-significance is reached, using the test statistic

$$\lambda_t = -(n - (p + q + 1)/2) \sum_{i=t}^{r} \ln(1 - \rho_i^2),$$

where n is the number of rows of $\mathbf{X}$, and $r = \min(p,q)$. The test statistic, λ_t, has approximately a chi-squared distribution with $(p + 1 - t)(q + 1 - t)$ degrees of freedom for large n.

12.2 Analysis of the decathlon data

The decathlon data were subjected to canonical correlation analysis, using the correlation matrix. The various matrices for the analysis are as follows.

$$\mathbf{S}_{11} = \begin{pmatrix} 1.0 & 0.67 & 0.30 & -0.46 \\ 0.67 & 1.0 & 0.10 & 0.06 \\ 0.30 & 0.10 & 1.0 & -0.30 \\ -0.46 & 0.06 & -0.30 & 1.0 \end{pmatrix}$$

$$\mathbf{S}_{22} = \begin{pmatrix} 1.0 & 0.46 & 0.18 & -0.04 & -0.18 & 0.52 \\ 0.46 & 1.0 & 0.23 & 0.29 & -0.13 & 0.10 \\ 0.18 & 0.23 & 1.0 & 0.38 & -0.13 & 0.03 \\ -0.04 & 0.29 & 0.38 & 1.0 & -0.35 & 0.04 \\ -0.18 & -0.13 & -0.13 & -0.35 & 1.0 & 0.04 \\ 0.52 & 0.10 & 0.03 & 0.04 & 0.04 & 1.0 \end{pmatrix}$$

$$\mathbf{S'_{12}} = \mathbf{S_{21}} = \begin{pmatrix} 0.70 & 0.55 & 0.08 & -0.42 \\ 0.38 & 0.06 & -0.01 & -0.34 \\ 0.17 & 0.20 & -0.03 & 0.20 \\ -0.52 & -0.38 & -0.38 & 0.30 \\ 0.11 & 0.05 & 0.34 & 0.22 \\ 0.27 & 0.18 & -0.21 & -0.10 \end{pmatrix}$$

$$\mathbf{S_{11}^{-1}S_{12}S_{22}^{-1}S_{21}} = \begin{pmatrix} 0.707 & 0.372 & 0.261 & -0.199 \\ 0.132 & 0.265 & 0.021 & -0.152 \\ 0.042 & 0.075 & 0.285 & 0.116 \\ -0.124 & -0.071 & 0.125 & 0.407 \end{pmatrix}$$

$$\mathbf{S_{22}^{-1}S_{21}S_{11}^{-1}S_{12}} = \begin{pmatrix} 0.403 & 0.065 & 0.066 & -0.360 & 0.053 & 0.060 \\ 0.188 & 0.270 & -0.009 & -0.091 & -0.024 & 0.130 \\ 0.132 & 0.044 & 0.157 & -0.118 & 0.171 & 0.067 \\ -0.489 & -0.282 & -0.065 & 0.360 & -0.029 & -0.172 \\ -0.118 & -0.085 & 0.123 & -0.076 & 0.337 & -0.112 \\ 0.027 & 0.109 & 0.012 & 0.128 & -0.139 & 0.135 \end{pmatrix}.$$

The non-zero eigenvalues of $\mathbf{S_{11}^{-1}S_{12}S_{22}^{-1}S_{21}}$ and $\mathbf{S_{22}^{-1}S_{21}S_{11}^{-1}S_{12}}$, are 0.884, 0.449, 0.201 and 0.130, giving the four canonical correlations as 0.94, 0.67, 0.45 and 0.36. The corresponding eigenvectors are

ρ	0.94	0.67	0.45	0.36
100 m	0.73	-0.60	1.60	0.12
400 m	0.21	0.37	-1.51	0.50
110 h	0.04	-0.75	-0.43	-0.63
1500 m	-0.21	-1.07	0.70	0.47

ρ	0.94	0.67	0.45	0.36
LJ	0.60	0.20	0.91	0.15
SP	0.28	-0.30	-0.98	-0.57
HJ	0.24	0.46	-0.24	0.64
DIS	-0.72	0.07	0.15	0.20
PV	-0.01	0.92	-0.12	-0.15
JAV	-0.06	-0.45	-0.73	0.55

The values of $\lambda_1, \lambda_2, \lambda_3$, and λ_4 are 35.86, 11.02, 4.16, and 1.59, on 24, 15, 8, and three degrees of freedom respectively. Here n is only 17 and hence the chi-squared approximation as to the distribution of λ_t must be questioned.

The first canonical variables are

$$U_1 = 0.73 \times 100\,\mathrm{m} + 0.21 \times 400\,\mathrm{m} + 0.04 \times 110\,\mathrm{h} - 0.21 \times 1500\,\mathrm{m}$$

$$V_1 = 0.60 \times \mathrm{LJ} + 0.28 \times \mathrm{SP} + 0.24 \times \mathrm{HJ} - 0.72 \times \mathrm{DIS} - 0.01 \times \mathrm{PV} - 0.06 \times \mathrm{JAV},$$

with canonical correlation coefficient 0.94. Interpretation of canonical variables using the coefficients requires some caution, since the variables comprising the canonical variables

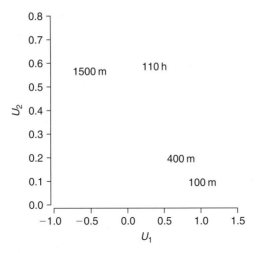

Figure 12.1 Coefficients of U_2 plotted against those for U_1

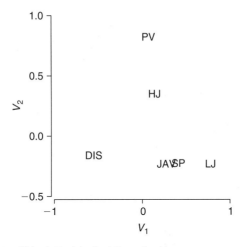

Figure 12.2 Coefficients of V_2 plotted against those for V_1

can be highly correlated with each other. Interpretation may be helped by looking at the correlations of the canonical variables with each of the original variables. The correlations of the four track canonical variables with the four track event are

	U_1	U_2	U_3	U_4
100 m	0.99	0.08	0.14	0.06
400 m	0.70	0.18	−0.43	0.55
110 h	0.34	0.57	−0.32	−0.68
1500 m	−0.55	0.55	0.00	0.63

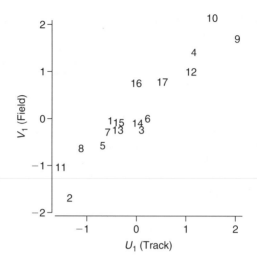

Figure 12.3 Scores for the first canonical correlation variables

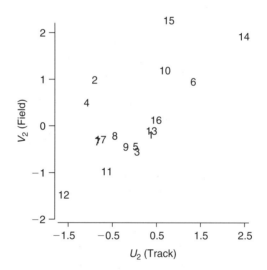

Figure 12.4 Scores for the second canonical correlation variables

The correlations of the four field canonical variables with the six field events are

	V_1	V_2	V_3	V_4
LJ	0.77	−0.26	−0.06	0.31
SP	0.39	−0.25	0.63	−0.22
HJ	0.13	0.32	0.25	0.65
DIS	−0.57	−0.19	0.24	0.35
PV	0.06	0.79	0.26	−0.20
JAV	0.26	−0.26	0.37	0.58

The main component of U_1 is the 100 m, but it also contrasts the 100 m, 400 m and 110 h with the 1500 m. The main component of V_1 is the LJ, and also, in contrast, the discus.

Figure 12.1 plots the coefficients for U_1 against those for U_2, and Figure 12.2 the coefficients for V_1 against those for V_2. These plots help to see which events are closely associated. The 100 metres and the 400 metres races are close together for the track events. For the field events, the javelin and shot are very close together, and the pole vault and high jump are reasonably close.

Figure 12.3 plots the values of U_1 and V_1 for the athletes, with labels as the rank order in which they finished the decathlon. The strong correlation between the variables can be seen. Not surprisingly, the order of the athletes starting at the bottom left of the plot, and going towards the upper right, well matches the order in which the athletes finished in the 100 m.

The second canonical correlation variables are dominated by the 110 h, 1500 m and the PV. Scores are plotted in Figure 12.4, where it can be seen that there are no athletes who score high on U_2 and low on V_2.

12.3 Exercises

1. The covariance matrix for the blood pressure data described in Chapter 3, is

$$\begin{pmatrix} X_1 & X_2 & Y_1 & Y_2 \\ 422.9 & 143.2 & 370.8 & 105.1 \\ 143.2 & 109.7 & 153.1 & 96.5 \\ 370.8 & 153.1 & 400.1 & 166.3 \\ 105.1 & 96.5 & 166.3 & 157.6 \end{pmatrix}.$$

The variables are Systolic-pre (X_1), Diastolic-pre (X_2), Systolic-post (Y_1) and Diastolic-post (Y_2), and are split into pre- and post-measurements. Find the canonical correlations and canonical correlation variables. Find the correlations of the original variables with the canonical correlation variables.

2. Show that, in general, the canonical correlation variables based on the correlation matrix are identical to those based on the covariance matrix.

3. The covariance matrix for the four variables X_1, X_2, Y_1, Y_2 is

$$\begin{pmatrix} 1 & \rho & \gamma_1 & \gamma_1 \\ \rho & 1 & \gamma_2 & \gamma_2 \\ \gamma_1 & \gamma_2 & 1 & \rho \\ \gamma_1 & \gamma_2 & \rho & 1 \end{pmatrix}.$$

Find the canonical correlations and the canonical correlation variables. Investigate how these depend upon ρ, γ_1, and γ_2.

4. Download the diabetes dataset from http://biostat.mc.vanderbilt.edu/twiki/bin/view/Main/DataSets and carry out a canonical correlation analysis. Use *chol, stab.glu, hdl, ratio* and *glyhb* for the X-variables, and *pb.1s, bp.1d, bp.2s, bp.2d, waist* and *hip* for the Y-variables.

13

Discriminant analysis and canonical variates analysis

Suppose you present yourself to a physician with an ankle that is swollen, bruised, and very painful, having fallen awkwardly playing basketball. The physician will assess the symptoms: how swollen is the ankle, can the patient walk on it, can the patient move the joint, can the patient move their toes, how much pain does the patient appear to be suffering, and so on. Based on these symptoms, the physician will make a decision: (i) the patient has suffered a badly sprained ankle, (ii) the patient has broken their ankle, or (iii) further information is required in the form of an X-ray. Based on the data about the ankle, the physician will place you in one of two groups of patients: (i) those who have sprained their ankle, or (ii) those who have broken their ankle, although in this case, he has the option to defer the choice until further information is available.

Another example from medicine is to place men with symptoms relating to the prostate into one of two groups: (i) those with prostate cancer, and (ii) those without prostate cancer. Data from digital rectal screening (DRE) and a blood test measuring the serum tumour marker prostate specific antigen (PSA) will be used by the physician to place a patient into one of the two groups. Those placed in group (i) would then be subjected to further tests.

Placing patients (or people, objects, observations) into one of two or more groups using data about the patients is called discriminant analysis. Usually, data are collected on patients of known group, the *training data*, and then the patient of unknown group is allocated to one of the groups by comparing their data to those for the known groups. There are various methods for allocating the patient to one of the groups, and here we explore just some of them. For a more extensive coverage, the reader is referred to McLachlan (1992).

Another use of discriminant analysis is the exploration of the differences between groups. So, for instance, we are interested in which variables show the differences between the group of patients that have prostate cancer and the group of patients that do not have prostate cancer. *Canonical variates analysis* can be used to do this.

13.1 Discriminant analysis

We start with an illustrative example. Collett (2003) gives a small dataset of the erythrocytes sedimentation rate (ESR) of thirty-two patients. ESR is the rate at which red blood cells

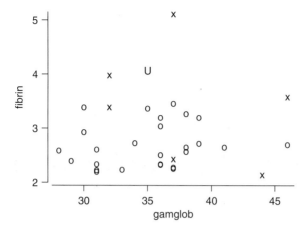

Figure 13.1 Scatterplot of fibrinogen against gamma globulin

(erythrocytes) settle out of suspension. Patients with a rate of twenty or above are considered unhealthy and have problems with rheumatic diseases, chronic infections and malignancies. Those patients with ESR below twenty are considered healthy. Two variables associated with ESR are fibrinogen and gamma gobulin levels. Is it possible to predict whether a patient is healthy, based on these two variables alone? Figure 13.1 shows a scatterplot for these two variables with the points labelled as healthy patient (O), or unhealthy patient (X). Let the group of healthy patients be known as Group 1, and the group of unhealthy patients, Group 2. Also shown in the scatterplot is a point (U) representing a patient of unknown group. To which group should we allocate this patient?

13.1.1 *K*-nearest neighbour discrimination

The nearest neighbour discriminant method allocates the observation of an unknown group to the group of that of its nearest neighbour. In order to be able to say which point is the nearest neighbour, a definition of distance between observations is needed. Referring to the discussion on proximity data of Chapter 1, there are several ways of measuring such distances. To be more robust against allocating to the wrong group, k nearest neighbours can be used, and the observation is allocated to the group of the majority of its k nearest neighbours. The value of k has to be chosen, as has the method of measuring distance between points. For the ESR data, we use Euclidean distance between the standardized observations, obtained by subtracting the sample mean, and dividing by the sample standard deviation. A plot of these can be seen in Figure 13.2. The observation of unknown origin, U, is closest to a Group 2 point, at a distance of 0.67, and hence would be allocated to the group of unhealthy patients. If k is chosen as five, then of the five nearest neighbours, three are from Group 1, and two from Group 2, and hence U would be allocated to Group 1. However, we might wish to compensate for the fact that there are twenty-six Group 1 points, and only six Group 2 points in our dataset, and hence favouring Group 1 much of the time. There are many variations of the k nearest neighbour method, see for example, Hastie et al. (2001).

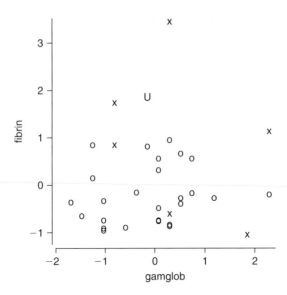

Figure 13.2 Scatterplot of fibrinogen against gamma globulin for the standardized observations

13.1.2 Maximum likelihood discrimination

The k-nearest neighbour method is a non-parametric discrimination method. We now look at a parametric method based on the likelihood. Suppose there are k groups, labelled G_1 to G_k. Let the variables used in the discrimination exercise be denoted by $\mathbf{x}$. Let the observations from group G_i come from a distribution with pdf (or probability function), $f_i(\mathbf{x})$. (Note that the distribution for each group is from the same family; e.g. all multivariate normal.) The likelihood, $L_i(\mathbf{x})$, for the ith group is equal to $f_i(\mathbf{x})$, but with emphasis on the parameters, rather than on $\mathbf{x}$. The *maximum likelihood discriminant rule* allocates an observation, $\mathbf{x}$, of unknown origin, to the group which gives rise to the largest likelihood. That is

$$\text{allocate } \mathbf{x} \text{ to } G_i, \text{ where } L_i(\mathbf{x}) = \max_j L_j(\mathbf{x}).$$

The two group multivariate normal case.   We look at the special case of two groups only, and where observations follow a multivariate normal distribution. Let $f_i(\mathbf{x})$ be the pdf of the multivariate normal distribution, $N_p(\boldsymbol{\mu}_i, \boldsymbol{\Sigma})$, $i = 1, 2$. Note that we are assuming a common variance matrix, $\boldsymbol{\Sigma}$. The likelihood, $L_i(\mathbf{x})$, is given by

$$L_i(\mathbf{x}) = (2\pi)^{-p/2}|\boldsymbol{\Sigma}|^{-1/2}\exp\left\{-\tfrac{1}{2}(\mathbf{x} - \boldsymbol{\mu}_i)'\boldsymbol{\Sigma}^{-1}(\mathbf{x} - \boldsymbol{\mu}_i)\right\}.$$

We allocate an observation, $\mathbf{x}$, of unknown origin to G_1 if

$$L_1(\mathbf{x}) > L_2(\mathbf{x}),$$

that is, if

$$(\mathbf{x} - \boldsymbol{\mu}_1)'\boldsymbol{\Sigma}^{-1}(\mathbf{x} - \boldsymbol{\mu}_1) < (\mathbf{x} - \boldsymbol{\mu}_2)'\boldsymbol{\Sigma}^{-1}(\mathbf{x} - \boldsymbol{\mu}_2), \tag{13.1}$$

after a small amount of rearranging. Expanding (13.1) and further rearranging, gives us the maximum likelihood discriminant rule

allocate $\mathbf{x}$ to G_1 if $(\boldsymbol{\mu}_2 - \boldsymbol{\mu}_1)'\boldsymbol{\Sigma}^{-1}\mathbf{x} - \frac{1}{2}(\boldsymbol{\mu}_2 - \boldsymbol{\mu}_1)'\boldsymbol{\Sigma}^{-1}(\boldsymbol{\mu}_1 + \boldsymbol{\mu}_2) < 0,$

allocate $\mathbf{x}$ to G_2 otherwise.

The term $h(\mathbf{x}) = (\boldsymbol{\mu}_2 - \boldsymbol{\mu}_1)'\boldsymbol{\Sigma}^{-1}\mathbf{x} - \frac{1}{2}(\boldsymbol{\mu}_2 - \boldsymbol{\mu}_1)'\boldsymbol{\Sigma}^{-1}(\boldsymbol{\mu}_1 + \boldsymbol{\mu}_2)$ is called *the discriminant*, and note that in this case, it is linear in $\mathbf{x}$. In practice, the population means $\boldsymbol{\mu}_1, \boldsymbol{\mu}_2$, and the population variance, $\boldsymbol{\Sigma}$, will not be known, and have to be estimated by their sample counterparts.

Example: The ESR data

The sample mean vectors and sample covariance matrices are

$$\bar{\mathbf{x}}_1 = \begin{pmatrix} 2.65 \\ 35.12 \end{pmatrix}, \quad \bar{\mathbf{x}}_2 = \begin{pmatrix} 3.39 \\ 38.00 \end{pmatrix},$$

$$S_1 = \begin{pmatrix} 0.165 & 0.288 \\ 0.288 & 17.466 \end{pmatrix}, \quad S_2 = \begin{pmatrix} 1.163 & -2.056 \\ -2.056 & 34.800 \end{pmatrix}.$$

The pooled estimate of $\boldsymbol{\Sigma}$ is

$$S_u = (25S_1 + 5S_2)/30 = \begin{pmatrix} 0.330 & -0.102 \\ -0.102 & 20.36 \end{pmatrix},$$

with inverse

$$S_u^{-1} = \begin{pmatrix} 3.027 & 0.015 \\ 0.015 & 0.049 \end{pmatrix}.$$

The linear discriminant and allocation rule is

allocate $\mathbf{x}$ to Group 1   if $2.277x_1 + 0.153x_2 - 12.473 < 0$

allocate $\mathbf{x}$ to Group 2   otherwise,

where x_1 is the fibrinogen measurement, and x_2 is the gamma globulin measurement.

Figure 13.3 shows the *discriminant boundary* within the sample space, where the two likelihoods are equal, i.e. $2.277x_1 + 0.153x_2 - 12.473 = 0$. An observation, $\mathbf{x}$, of unknown origin is allocated to Group 1 if it lies below the discriminant boundary, and to Group 2 if it lies above. The point U is thus allocated to Group 2. We can see that there is a Group 2 observation in the training data that lies within the Group 2 portion of the sample space, and likewise, there are many Group 1 observations that lie in the Group 2 region. Rarely would we find a discriminant boundary that completely separates the two groups of points.

When allocating observations of unknown group to one or other of the groups, we are bound to make some errors, allocating a Group 1 observation to Group 2, and vice versa. Depending on context, one of the two types of error, allocating a Group 1 to Group 2, or a Group 2 to Group 1, may be more important than the other. For instance, it is better to send

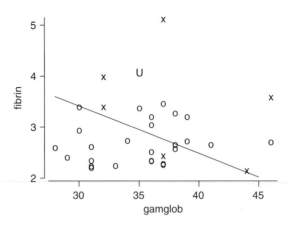

Figure 13.3 The discriminant boundary for the ESR data

a patient who does not actually have a critical disease for further tests, than to tell a patient who does have the disease, that they are healthy and free from the disease. The discriminant boundary can be adjusted appropriately for this.

Using prior knowledge.   Suppose the prior probability that an observation of unknown origin belongs to Group i, is π_i, $i = 1, 2$ $(\pi_1 + \pi_2 = 1)$. Then the allocation rule is now

allocate **x** to Group 1   if $\pi_1 L_1(\mathbf{x}) > \pi_2 L_2(\mathbf{x})$

allocate **x** to Group 2   otherwise.

The discriminant rule becomes

allocate **x** to Group 1   if

$$(\boldsymbol{\mu}_2 - \boldsymbol{\mu}_1)'\boldsymbol{\Sigma}^{-1}\mathbf{x} - \tfrac{1}{2}(\boldsymbol{\mu}_2 - \boldsymbol{\mu}_1)'\boldsymbol{\Sigma}^{-1}(\boldsymbol{\mu}_1 + \boldsymbol{\mu}_2) < \log{(\pi_1/\pi_2)},$$

allocate **x** to Group 2   otherwise.

We can see that varying the ratio of π_1/π_2 will raise or lower the discriminant boundary in Figure 13.3, but keeping it parallel to the original, and when $\pi_1 = \pi_2 = 0.5$, we actually have our original likelihood discriminant rule. The simplest choice of π_i is to equate it to an estimate of the proportion of the population that fall into Group i. If the training data are a random sample from the population, then the proportion of observations falling into group i can be used as an estimate of π_i. On the other hand, if the training data have been collected in some other way, e.g. a set number of observations per group, then alternative data or methods must be used to estimate π_i. It is also possible to incorporate costs of making allocation errors (see for example Mardia et al., 1979), and these again alter the position of the discriminant boundary.

For the ESR data, the proportion of observations in Group 1 is 26/32, and for Group 2, 6/32. Hence, we might estimate π_1/π_2 as 4.33, and hence the discriminant rule becomes

allocate **x** to Group 1   if $2.277x_1 + 0.153x_2 - 13.94 < 0$

allocate **x** to Group 2   otherwise,

which will raise the boundary relative to the y-axis, making less allocation errors for Group 1, but more for Group 2.

More than two variables.   When **x** is only two-dimensional, a scatterplot of the training data, together with the discriminant boundary, clearly illustrates the partitioning of the sample space, group membership, and how well the discrimination technique is performing. Indeed, it could be argued that no theoretical method is needed for the discrimination, and that the analyst could draw a straight line between the groups by eye, and use that as the discriminant boundary. Unfortunately, we lose this visualization in higher dimensions. The discriminant boundary becomes a hyperplane in p-dimensional space.

More than two groups.   Data on polychaetes, courtesy of Jeff Dodgson, Napier University (http://www.maths.napier.ac.uk/~jeff/courses/mmmult.html), are used for illustration. Polychaetes include over 10,000 species of worm, including types of sand worm, tube worm, clam worm, and others. The dataset has sixty-seven specimens from four species. The variables measured were: x_1 – _area_, x_2 – _perimeter_, x_3 – _elongation_, x_4 – _roundness_, and x_5 – _feret_. For convenience, the data were standardized so that each variable had overall mean zero and variance unity.

Assuming a multivariate normal distribution, from equation (13.1), the maximum likelihood discriminant for the ith species is

$$h_i(\mathbf{x}) = \boldsymbol{\mu}_i'\boldsymbol{\Sigma}^{-1}\mathbf{x} - \tfrac{1}{2}\boldsymbol{\mu}_i'\boldsymbol{\Sigma}^{-1}\boldsymbol{\mu}_i,$$

which is estimated as

$$\bar{\mathbf{x}}_i'\mathbf{S}_u^{-1}\mathbf{x} - \tfrac{1}{2}\bar{\mathbf{x}}_i'\mathbf{S}_u^{-1}\bar{\mathbf{x}}_i'.$$

The sample mean vectors and the pooled estimate of $\boldsymbol{\Sigma}$ are

$$\bar{\mathbf{x}}_1 = (-0.42, -0.12, 0.66, -0.68, -0.39)'$$
$$\bar{\mathbf{x}}_2 = (-0.28, -0.22, 0.18, -0.02, -0.20)'$$
$$\bar{\mathbf{x}}_3 = (-0.81, -1.09, -0.39, -0.23, -0.96)'$$
$$\bar{\mathbf{x}}_4 = (1.43, 1.33, -0.49, 0.87, 1.44)'$$

$$\mathbf{S}_u = \begin{bmatrix} 0.27 & 0.19 & -0.20 & 0.20 & 0.23 \\ 0.19 & 0.27 & 0.07 & -0.11 & 0.18 \\ -0.20 & 0.07 & 0.83 & -0.60 & -0.20 \\ 0.20 & -0.11 & -0.60 & 0.71 & 0.18 \\ 0.23 & 0.18 & -0.20 & 0.18 & 0.22 \end{bmatrix}.$$

The discriminants are then

$$h_1(\mathbf{x}) = 0.87x_1 + 4.67x_2 - 0.50x_3 + 1.10x_4 - 7.88x_5 - 0.55$$
$$h_2(\mathbf{x}) = -4.60x_1 - 3.29x_2 + 1.03x_3 - 0.44x_4 + 8.02x_5 - 0.33$$
$$h_3(\mathbf{x}) = 16.55x_1 + 10.14x_2 - 5.02x_3 + 1.74x_4 - 36.53x_5 - 6.08$$
$$h_4(\mathbf{x}) = -10.29x_1 - 9.67x_2 + 3.75x_3 - 2.07x_4 + 30.68x_5 - 6.52,$$

and a specimen, $\mathbf{x}^*$, of unknown species is allocated to the species which has the largest value of $h_i(\mathbf{x}^*)$. For example, the values of $h_i(\mathbf{x}^*)$, for a polychaete with standardized measurements $(-0.88, -1.10, 0.37, -0.73, -1.07)$ are: $h_1(\mathbf{x}^*)=0.99$, $h_2(\mathbf{x}^*)=-0.54$, $h_3(\mathbf{x}^*)=4.17$ and $h_1(\mathbf{x}^*)=-16.78$. Thus this polychaete would be classified as species 3. Mahalanobis squared distances between species are calculated as

$$d_{ij}^2 = (\bar{\mathbf{x}}_i - \bar{\mathbf{x}}_j)'\mathbf{S}_u^{-1}(\bar{\mathbf{x}}_i - \bar{\mathbf{x}}_j),$$

and these are

$$\begin{bmatrix} 0 & 1.46 & 9.87 & 19.57 \\ 1.46 & 0 & 14.17 & 14.23 \\ 9.87 & 14.17 & 0 & 48.32 \\ 19.57 & 14.23 & 48.32 & 0 \end{bmatrix}.$$

It can be seen that species 1 and 2 are closest, while species 3 and 4 are furthest apart.

The following table gives the results of allocating the sixty-seven specimens to species group according to the largest value of $h_i(\mathbf{x})$.

From species	Classified into species			
1	12	4	0	0
2	5	13	1	0
3	1	0	14	0
4	0	0	0	17

The proportion of correct classifications is 0.84. This is optimistically high, since the data being classified have also estimated the discriminants. If *cross validation* is used, where, in turn, one observation is left out from the estimation of the discriminants, which are then used to classify that particular observation, the proportion of correct classifications reduces to 0.75.

13.1.3 Probabilities of misclassification

We consider the case of two groups. The discriminant rule is allocate $\mathbf{x}$ to Group 1 if

$$h(\mathbf{x}) = (\boldsymbol{\mu}_1 - \boldsymbol{\mu}_2)'\boldsymbol{\Sigma}^{-1}\left(\mathbf{x} - \tfrac{1}{2}(\boldsymbol{\mu}_1 + \boldsymbol{\mu}_2)\right) < 0.$$

For the two types of misclassification, let

$$p(G_1|G_2) = \Pr(\mathbf{x} \text{ is allocated to Group 1 given } \mathbf{x} \text{ is from Group 2})$$
$$p(G_2|G_1) = \Pr(\mathbf{x} \text{ is allocated to Group 2 given } \mathbf{x} \text{ is from Group 1}).$$

Concentrating upon the first of these misclassification probabilities,

$$p(G_1|G_2) = \Pr\{h(\mathbf{x}) < 0|G_2\}$$
$$= \Pr\left\{(\boldsymbol{\mu}_2 - \boldsymbol{\mu}_1)'\boldsymbol{\Sigma}^{-1}(\mathbf{x} - \tfrac{1}{2}(\boldsymbol{\mu}_1 + \boldsymbol{\mu}_2)) < 0 | \mathbf{x} \sim N_p(\boldsymbol{\mu}_2, \boldsymbol{\Sigma})\right\}$$
$$= \Pr\left\{(\boldsymbol{\mu}_2 - \boldsymbol{\mu}_1)'\boldsymbol{\Sigma}^{-1}\mathbf{x} < \tfrac{1}{2}(\boldsymbol{\mu}_2 - \boldsymbol{\mu}_1)'\boldsymbol{\Sigma}^{-1}(\boldsymbol{\mu}_1 + \boldsymbol{\mu}_2) | \mathbf{x} \sim N_p(\boldsymbol{\mu}_2, \boldsymbol{\Sigma})\right\}$$

Now $(\boldsymbol{\mu}_2 - \boldsymbol{\mu}_1)'\boldsymbol{\Sigma}^{-1}\mathbf{x}$ has a univariate normal distribution with mean and variance given by

$$E\left((\boldsymbol{\mu}_2 - \boldsymbol{\mu}_1)'\boldsymbol{\Sigma}^{-1}\mathbf{x}\right) = (\boldsymbol{\mu}_2 - \boldsymbol{\mu}_1)'\boldsymbol{\Sigma}^{-1}\boldsymbol{\mu}_2$$
$$\mathrm{var}\left((\boldsymbol{\mu}_2 - \boldsymbol{\mu}_1)'\boldsymbol{\Sigma}^{-1}\mathbf{x}\right) = (\boldsymbol{\mu}_2 - \boldsymbol{\mu}_1)'\boldsymbol{\Sigma}^{-1}(\boldsymbol{\mu}_2 - \boldsymbol{\mu}_1).$$

Hence

$$
\begin{aligned}
p(G_1|G_2) &= \Phi\left(\frac{\frac{1}{2}(\boldsymbol{\mu}_2 - \boldsymbol{\mu}_1)'\boldsymbol{\Sigma}^{-1}(\boldsymbol{\mu}_1 + \boldsymbol{\mu}_2) - (\boldsymbol{\mu}_2 - \boldsymbol{\mu}_1)'\boldsymbol{\Sigma}^{-1}\boldsymbol{\mu}_2}{\{(\boldsymbol{\mu}_2 - \boldsymbol{\mu}_1)'\boldsymbol{\Sigma}^{-1}(\boldsymbol{\mu}_2 - \boldsymbol{\mu}_1)\}^{1/2}}\right) \\
&= \Phi\left(\frac{\frac{1}{2}(\boldsymbol{\mu}_2 - \boldsymbol{\mu}_1)'\boldsymbol{\Sigma}^{-1}(\boldsymbol{\mu}_1 - \boldsymbol{\mu}_2)}{\{(\boldsymbol{\mu}_2 - \boldsymbol{\mu}_1)'\boldsymbol{\Sigma}^{-1}(\boldsymbol{\mu}_2 - \boldsymbol{\mu}_1)\}^{1/2}}\right) \\
&= \Phi\left(-\tfrac{1}{2}\{(\boldsymbol{\mu}_2 - \boldsymbol{\mu}_1)'\boldsymbol{\Sigma}^{-1}(\boldsymbol{\mu}_2 - \boldsymbol{\mu}_1)\}^{1/2}\right),
\end{aligned}
$$

and a similar calculation shows that $p(G_2|G_1) = p(G_1|G_2)$.

For the ESR data, $-\frac{1}{2}\{(\bar{\mathbf{x}}_2 - \bar{\mathbf{x}}_1)'\mathbf{S}_u^{-1}(\bar{\mathbf{x}}_2 - \bar{\mathbf{x}}_1)\}^{1/2} = 1.061$, and hence an estimate of the probability of misclassification is 0.14.

13.2 Quadratic discrimination

The likelihood discriminant rule obtained from the multivariate normal distribution was linear in the variables, x_i. This was a consequence of assuming a common covariance matrix, $\boldsymbol{\Sigma}$, for each of the groups. If this condition is relaxed, we obtain a *quadratic discriminant function*. We consider the two-group case only, where the covariance matrices are $\boldsymbol{\Sigma}_1$ and $\boldsymbol{\Sigma}_2$. Following the previous derivation for the linear discriminant, the discriminant rule in (13.1) is to allocate an observation, $\mathbf{x}$, of unknown origin to Group 1, if

$$\tfrac{1}{2}\log|\boldsymbol{\Sigma}_1| + (\mathbf{x} - \boldsymbol{\mu}_1)'\boldsymbol{\Sigma}_1^{-1}(\mathbf{x} - \boldsymbol{\mu}_1) < \tfrac{1}{2}\log|\boldsymbol{\Sigma}_2| + (\mathbf{x} - \boldsymbol{\mu}_2)'\boldsymbol{\Sigma}_2^{-1}(\mathbf{x} - \boldsymbol{\mu}_2).$$

Rearranging this expression, the discriminant rule is

allocate $\mathbf{x}$ to Group 1 if

$$
\begin{aligned}
h(\mathbf{x}) = {}& \mathbf{x}'\left(\boldsymbol{\Sigma}_1^{-1} - \boldsymbol{\Sigma}_2^{-1}\right)\mathbf{x} - 2\left(\boldsymbol{\mu}_1'\boldsymbol{\Sigma}_1^{-1} + \boldsymbol{\mu}_2'\boldsymbol{\Sigma}_2^{-1}\right)\mathbf{x} + \left(\boldsymbol{\mu}_1'\boldsymbol{\Sigma}_1^{-1}\boldsymbol{\mu}_1 - \boldsymbol{\mu}_2'\boldsymbol{\Sigma}_2^{-1}\boldsymbol{\mu}_2\right) \\
& + \frac{1}{2}\log\left(|\boldsymbol{\Sigma}_1|/|\boldsymbol{\Sigma}_2|\right) < 0
\end{aligned}
$$

allocate $\mathbf{x}$ to Group 2 otherwise.

The discriminant contains the quadratic form, $\mathbf{x}'\left(\boldsymbol{\Sigma}_1^{-1} - \boldsymbol{\Sigma}_2^{-1}\right)\mathbf{x}$, and thus has quadratic terms. This gives rise to a quadratic discriminant boundary.

13.3 Non-parametric discrimination

In this section we look at some discrimination methods that do not make any assumption about the distribution of $\mathbf{x}$. We start with the classic method due to Fisher.

13.3.1 Fisher's linear discriminant function

First, we consider the two-group case. Let $x_{i1}, \ldots, x_{in_i}$ be a sample from group G_i, for $i = 1, 2$. Our aim is to find a constant vector, $\mathbf{a}$, so that the univariate random variable, given by $Y = \mathbf{a}'\mathbf{x}$, is a 'good' discriminant. That is, for one group we would have the y values less than zero, say, and for the other group, the y values greater than zero. The y values for the two groups would be pushed as far apart as possible.

If the mean vectors for $\mathbf{x}$ for the two groups were $\boldsymbol{\mu}_1$ and $\boldsymbol{\mu}_2$, and if we were testing the null hypothesis $H_0 : \mathbf{a}'\boldsymbol{\mu}_1 = \mathbf{a}'\boldsymbol{\mu}_2$, then we would use the t-statistic

$$\frac{|\bar{y}_2 - \bar{y}_1|}{\{s^2(n_1^{-1} + n_2^{-1})\}^{1/2}}.$$

Squaring and substituting,

$$t^2 = \frac{\{\mathbf{a}'(\bar{\mathbf{x}}_2 - \bar{\mathbf{x}}_1)\}^2}{(\mathbf{a}'\mathbf{S}\mathbf{a})(n_1^{-1} + n_2^{-1})},$$

where $\mathbf{S} = \{(n_1 - 1)\mathbf{S}_1 + (n_2 - 1)\mathbf{S}_2\}/(n_1 + n_2 - 2)$.

Although we are not testing hypotheses, but are discriminating one group from another, choosing $\mathbf{a}$ to maximize t^2 will separate or discriminate between the two groups means. Fisher's linear discriminant does just this. Now

$$\frac{\partial t^2(\mathbf{a})}{\partial \mathbf{a}} = \frac{2(\mathbf{a}'\mathbf{S}\mathbf{a})\{\mathbf{a}'(\bar{\mathbf{x}}_2 - \bar{\mathbf{x}}_1)\}(\bar{\mathbf{x}}_2 - \bar{\mathbf{x}}_1) - 2\{\mathbf{a}'(\bar{\mathbf{x}}_2 - \bar{\mathbf{x}}_1)\}^2 \mathbf{S}\mathbf{a}}{(\mathbf{a}'\mathbf{S}\mathbf{a})^2(n_1^{-1} + n_2^{-1})}.$$

Equating to $\mathbf{0}$, and rearranging gives

$$\mathbf{a} = \frac{\mathbf{a}'\mathbf{S}\mathbf{a}}{\mathbf{a}'(\bar{\mathbf{x}}_2 - \bar{\mathbf{x}}_1)} \mathbf{S}^{-1}(\bar{\mathbf{x}}_2 - \bar{\mathbf{x}}_1),$$

but since $\mathbf{a}$ can have an arbitrary scaling factor, we use $\mathbf{a} = \mathbf{S}^{-1}(\bar{\mathbf{x}}_2 - \bar{\mathbf{x}}_1)$, and Fisher's linear discriminant, $h(\mathbf{x}) = \mathbf{a}'\mathbf{x}$, is then

$$h(\mathbf{x}) = (\bar{\mathbf{x}}_2 - \bar{\mathbf{x}}_1)'\mathbf{S}^{-1}\mathbf{x}.$$

Note that this is the same discriminant rule as that for maximum likelihood discrimination for two multivariate normal distributions.

To allocate a new observation, $\mathbf{x}^*$, to one of the two groups, calculate

$$y = (\bar{\mathbf{x}}_2 - \bar{\mathbf{x}}_1)'\mathbf{S}^{-1}\mathbf{x}^*$$
$$\bar{y}_1 = (\bar{\mathbf{x}}_2 - \bar{\mathbf{x}}_1)'\mathbf{S}^{-1}\bar{\mathbf{x}}_1$$
$$\bar{y}_2 = (\bar{\mathbf{x}}_2 - \bar{\mathbf{x}}_1)'\mathbf{S}^{-1}\bar{\mathbf{x}}_2.$$

If y is closer to $\bar{y}_1$ than $\bar{y}_2$, then allocate $\mathbf{x}^*$ to Group G_1, and otherwise to group G_2. The discriminant boundary, where y is as close to $\bar{y}_1$ as it is to $\bar{y}_2$, is given by

$$(\bar{\mathbf{x}}_2 - \bar{\mathbf{x}}_1)'\mathbf{S}^{-1}\mathbf{x} = \tfrac{1}{2}(\bar{y}_1 + \bar{y}_2),$$

i.e.

$$(\bar{\mathbf{x}}_2 - \bar{\mathbf{x}}_1)'\mathbf{S}^{-1}(\mathbf{x} - \tfrac{1}{2}(\bar{\mathbf{x}}_1 + \bar{\mathbf{x}}_2)) = 0.$$

The same idea can be applied to the case of three or more groups, where the quantity to be maximized is the ratio of the between-groups sum of squares to the within-group sum of squares for the y's. Let our data be $\mathbf{x}_{ij}$, $(i = 1, \ldots, g; \; j = 1, \ldots, n_i)$, where we have g groups, $G_1, \ldots, G_g$. For a particular, $\mathbf{a}$, the transformed $\mathbf{x}$ values for each group are y_{ij}, using $y = \mathbf{a}'\mathbf{x}$. The between-groups sum of squares calculated for the y's is

$$\sum_i n_i(\bar{y}_i - \bar{y})^2 = \sum_i n_i\{\mathbf{a}'(\bar{\mathbf{x}}_i - \bar{\mathbf{x}})\}^2 = \mathbf{a}'\left\{\sum_i n_i(\bar{\mathbf{x}}_i - \bar{\mathbf{x}})(\bar{\mathbf{x}}_i - \bar{\mathbf{x}})'\right\}\mathbf{a} = \mathbf{a}'\mathbf{Ba},$$

and the within-groups sum of squares is

$$\sum_i \sum_j (y_{ij} - \bar{y}_i)^2 = \mathbf{a}'\left\{\sum_i \sum_j (\mathbf{x}_{ij} - \bar{\mathbf{x}}_i)(\mathbf{x}_i - \bar{\mathbf{x}}_i)'\right\}\mathbf{a} = \mathbf{a}'\mathbf{Wa}.$$

The vector $\mathbf{a}$ is found that maximizes

$$R = \frac{\mathbf{a}'\mathbf{Ba}}{\mathbf{a}'\mathbf{Wa}}. \tag{13.2}$$

Now since scaling $\mathbf{a}$ by a factor will not change the value of R, we can choose the denominator in (13.2) to be equal to a particular value, and in our case unity, i.e. $\mathbf{a}'\mathbf{Wa} = 1$. We now maximize $R = \mathbf{a}'\mathbf{Ba}$ subject to the constraint $\mathbf{a}'\mathbf{Wa} = 1$, using a Lagrange multiplier. Let

$$R = \mathbf{a}'\mathbf{Ba} - \lambda(\mathbf{a}'\mathbf{Wa} - 1),$$

and then differentiating R and equating to $\mathbf{0}$ gives

$$\frac{\partial R}{\partial \mathbf{a}} = 2\mathbf{Ba} - 2\lambda\mathbf{Wa} = \mathbf{0}. \tag{13.3}$$

Hence

$$(\mathbf{W}^{-1}\mathbf{B} - \lambda\mathbf{I})\mathbf{a} = \mathbf{0},$$

and thus λ is an eigenvalue of $\mathbf{W}^{-1}\mathbf{B}$. From (13.3), we have $\mathbf{Ba} = \lambda\mathbf{Wa}$, and therefore the maximum value of R is given by $R_{\max} = \mathbf{a}'\mathbf{Ba} = \mathbf{a}'\lambda\mathbf{Wa} = \lambda$. Thus the eigenvalue of $\mathbf{W}^{-1}\mathbf{B}$ giving $R_{\max}$ is the largest eigenvalue, and $\mathbf{a}$ is the corresponding eigenvector.

Once Fisher's linear discriminant, $\mathbf{a}'\mathbf{x}$, has been found, an observation, $\mathbf{x}^*$, of unknown group is allocated to one of the groups according to the value of $y^* = \mathbf{a}'\mathbf{x}^*$. From the training data, group means, $\bar{y}_i = \mathbf{a}'\bar{\mathbf{x}}_i$, $(i = 1, \ldots, g)$ are found, and then the distance of y^* to each $\bar{y}_i$. The observation $\mathbf{x}^*$ is then allocated to the group which has the smallest of these distances.

Example

For the polychaete data the between-groups and within-groups sums of squares matrices are

$$
\mathbf{B} = \begin{bmatrix}
48.76 & 47.43 & -12.59 & 28.65 & 50.27 \\
47.43 & 49.06 & -6.66 & 24.79 & 49.81 \\
-12.59 & -6.66 & 13.93 & -13.22 & -11.22 \\
28.65 & 24.79 & -13.22 & 21.19 & 29.02 \\
50.27 & 49.81 & -11.22 & 29.02 & 52.28
\end{bmatrix},
$$

$$
\mathbf{W} = \begin{bmatrix}
17.24 & 11.83 & -12.43 & 12.30 & 14.78 \\
11.83 & 16.94 & 4.16 & -6.86 & 11.05 \\
-12.43 & 4.16 & 52.07 & -37.78 & -12.77 \\
12.30 & -6.86 & -37.78 & 44.81 & 11.39 \\
14.78 & 11.05 & -12.77 & 11.39 & 13.72
\end{bmatrix}.
$$

The maximum eigenvalue of $\mathbf{W}^{-1}\mathbf{B}$ is 6.263 and has eigenvector $(-0.362, -0.282, 0.120, -0.055, 0.932)'$. This leads to Fisher's discriminant

$$
h(\mathbf{x}) = -0.362x_1 - 0.282x_2 + 0.120x_3 - 0.055x_4 + 0.932x_5.
$$

The values of $\mathbf{a}'\bar{\mathbf{x}}_i$ are: $\mathbf{a}'\bar{\mathbf{x}}_1 = -0.064$, $\mathbf{a}'\bar{\mathbf{x}}_2 = 0.004$, $\mathbf{a}'\bar{\mathbf{x}}_3 = -0.328$ and $\mathbf{a}'\bar{\mathbf{x}}_4 = 0.345$. For the polychaete of unknown species with measurements $(-0.88, -1.10, 0.37, -0.73, -1.07)$, $h(\mathbf{x}) = -0.284$, and hence would be allocated to species 3.

13.3.2 Kernel density discrimination

For maximum likelihood discrimination, we need to know the form of the probability density function for the random vectors in order to write down the likelihood. An alternative approach is to estimate the pdf using a kernel density estimator. For a univariate random variable, the kernel density estimator takes the form

$$
\hat{f}(x) = \frac{1}{N} \sum_{i=1}^{N} K(x, x_i; \lambda),
$$

where K is a *kernel* function. A popular choice of K is the *Gaussian kernel*,

$$
K(x, x_i; \lambda) = \frac{1}{\sqrt{2\pi\lambda}} \exp\left\{ -\tfrac{1}{2}(x - x_i)^2/\lambda \right\},
$$

which is a normal probability density function with mean x_i and variance λ. Essentially, a normal density function is placed at each of the data points, x_i, and then these density functions are summed. This gives an estimate of the density function for our data. To draw the estimate, many values of x are chosen that cover the range of the data, and then $\hat{f}(x)$ is calculated for each. The value of λ, which is called the *bandwidth*, has to be chosen. Results can be sensitive to the value of the bandwidth. See Silverman (1986) for further details.

For multivariate data, a *p-dimensional* kernel has to be used, which might be possible for low values of p, but for large values of p, sparseness of data will cause problems. An alternative is to assume the variables are independent and estimate the joint density as a product of univariate densities.

For discrimination, a kernel density function estimate is found for each group, G_i, and then an observation, $\mathbf{x}^*$, of unknown group is allocated to group G_i, if group G_i has the largest value of the density at the point, $\mathbf{x}^*$, i.e.

$$\text{allocate } \mathbf{x} \text{ to } G_i, \quad \text{where } \hat{f}_i(\mathbf{x}^*) = \max_j \hat{f}_j(\mathbf{x}^*).$$

13.3.3 Logistic discrimination

Logistic discrimination is a partial distributional method, in which the full form of the distribution of $\mathbf{x}$ is not specified, but only the ratio of likelihoods. We only consider the two-group case, with groups G_1 and G_2. In logistic discrimination, the log of the ratio of the likelihoods is modelled as

$$\log\left\{\frac{L(\mathbf{x}|G_1)}{L(\mathbf{x}|G_2)}\right\} = \boldsymbol{\beta}'\mathbf{x}, \tag{13.4}$$

where $\boldsymbol{\beta}' = (\beta_0, \beta_1, \ldots, \beta_p)$ is a vector of parameters, and $\mathbf{x}' = (x_0, x_1, \ldots, x_p)$, with $x_0 \equiv 1$. Several common distributions share this relationship for their likelihoods. Logistic discrimination is very useful for binary and other categorical data, and also for mixed data where some variables are continuous and others categorical. In practice, the β's need to be estimated from training data.

From equation (13.4), and writing the likelihoods as $\Pr(G_i|\mathbf{x})$,

$$\frac{\Pr(G_1|\mathbf{x})}{\Pr(G_2|\mathbf{x})} = \exp(\boldsymbol{\beta}'\mathbf{x}).$$

Replacing $\Pr(G_2|\mathbf{x})$ by $1 - \Pr(G_1|\mathbf{x})$, and after some rearranging,

$$\Pr(G_1|\mathbf{x}) = \frac{e^{\boldsymbol{\beta}'\mathbf{x}}}{1 + e^{\boldsymbol{\beta}'\mathbf{x}}} \tag{13.5}$$

$$\Pr(G_2|\mathbf{x}) = \frac{1}{1 + e^{\boldsymbol{\beta}'\mathbf{x}}}. \tag{13.6}$$

If, in the population at large, the ratio of individuals in the two groups is in the ratio $\pi_1 : \pi_2$, or prior probabilities for the two groups are π_1 and π_2, then by Bayes' theorem

$$L(\mathbf{x}|G_i) = \frac{\Pr(G_i|\mathbf{x})p(\mathbf{x})}{\Pr(G_i)},$$

which now gives the ratio of group membership as

$$\frac{Pr(G_1|\mathbf{x})\pi_2}{Pr(G_2|\mathbf{x})\pi_1} = \exp(\boldsymbol{\beta}'\mathbf{x}),$$

which then leads to the same group membership probabilities (13.5) and (13.6), but with β_0 replaced by $\beta_0^* = \beta_0 + \log(\pi_1/\pi_2)$.

The estimation of β_0^* can be affected by the method by which the training data are obtained. There are three sampling scenarios: (i) *mixture sampling* – where observations follow a joint distribution for $(G, \mathbf{X})$, and then the proportion of sample points from group G_i is an estimate of π_i; (ii) *x-conditional sampling* – where $\mathbf{x}$ is fixed and the group membership G_i is noted; (iii) *separate sampling* – x is sampled from each group with a fixed number of sample points from each.

It can be shown that, in each case, the same maximum likelihood estimate of $\boldsymbol{\beta}$ is obtained, but that for separate sampling an estimate of $\log(\pi_1/\pi_2)$ would need to be supplied, as it cannot be estimated from the training data.

Essentially, the likelihood for all three sampling schemes is

$$L = \prod_{r=1}^{N}\{Pr(G_1|\mathbf{x}_r)\}^{n_{r1}}\{Pr(G_2|\mathbf{x}_r)\}^{n_{r2}},$$

where N is the total size of the training data, and $n_{ri} = 1$ if the rth point of the training data is in group i, and zero otherwise.

The log-likelihood, l, is

$$l = \sum_{r=1}^{N}\{n_{r1} \log p_1(\mathbf{x}_r) + n_{r2} \log p_2(\mathbf{x}_r)\},$$

writing $p_i(\mathbf{x}) = Pr(G_i|\mathbf{x})$ for convenience. Differentiating l with respect to β_k, and equating to zero, gives the likelihood equations

$$\sum_{r=1}^{N}\{n_{r1}p_2(\mathbf{x}_r) - n_{r2}p_1(\mathbf{x}_r)\}x_{rk} = 0 \quad (k = 0, 1, \dots, p).$$

These have to be solved numerically.

Example

For the ESR data, incorporating weights for the differing sized groups, the estimates of the β's and their estimated standard errors are:

$$\hat{\beta}_0 = 11.54, \quad \text{s.e.} = 1.59$$

$$\hat{\beta}_1 = -1.68, \quad \text{s.e.} = 0.24$$

$$\hat{\beta}_2 = -0.18, \quad \text{s.e.} = 0.03.$$

Thus the discriminant rule based on the logistic model is

$$\text{allocate } \mathbf{x} \text{ to } G_1, \quad \text{when } p_i(\mathbf{x}) = \frac{e^{11.54-1.68x_1-0.18x_2}}{1+e^{11.54-1.68x_1-0.18x_2}} > 0.5$$

$$\text{allocate to } G_2 \quad \text{otherwise.}$$

The discriminant boundary is given by $11.54 - 1.68x_1 - 0.18x_2 = 0$. Compare this to the boundary in Figure 13.3 for maximum likelihood discrimination using multivariate normal distributions. When the original data were reclassified according to the discriminant rule, three out of six Group 1 observations, and seven out of twenty-six Group 2 observations were misclassified. Again, this is only an illustrative dataset, and these results are not surprising.

13.4 Support vector machines

Support vector machines (SVM) are a popular discriminant tool used by data miners (see Chapter 18). They combine a quadratic programming approach to finding a linear discriminant function, with the adoption of kernel basis functions, which transform the discriminant away from linearity. See Cristianini and Shawe-Taylor (2000) or Hastie et al. (2001) for a full discussion. Here, we can only give a brief introduction to SVMs, and the following relies heavily on these two books. To set the scene, Figure 13.4 shows the training data for two groups, G_1 and G_2, labelled **o** and **x**, for two variables, x_1 and x_2.

The two groups are entirely separated. The aim is to find the best discriminant, $h(\mathbf{x})$, that splits the two groups, best being defined as the centre line of the widest rectangle that can be placed between the two sets of points. The dotted lines in the figure give this

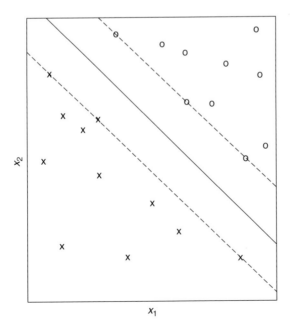

Figure 13.4 SVM discriminant for two variables

rectangle, and the solid line, the discriminant. With p-dimensional data, the discriminant and the boundaries would be hyperplanes. The points that lie on the boundary are called the *support vectors* as they are critical in defining the discriminant. If they are moved or deleted, then the discriminant changes. On the other hand, if the other points are moved, and as long as they do not become new support vectors, then the discriminant remains the same.

Now let $y_i = 1$ if the ith training point belongs to group G_1, and $y_i = -1$ if it belongs to group G_2. The discriminant can be written as $h(\mathbf{x}) = \beta_0 + \boldsymbol{\beta}'\mathbf{x} = 0$. A point $\mathbf{x}^*$ is allocated to group G_i if $h(\mathbf{x}^*) > 0$, and to group G_2 if $h(\mathbf{x}^*) < 0$. We actually change this allocation rule, by allocating to Group G_1 for points lying above the upper dotted line, and to Group G_2 for points lying below the lower dotted line. With an appropriate scaling of β_0 and $\boldsymbol{\beta}$, the allocation rule becomes

$$\text{allocate } \mathbf{x}^* \text{ to } G_1 \quad \text{if } \beta_0 + \boldsymbol{\beta}'\mathbf{x} \geq 1$$
$$\text{allocate } \mathbf{x}^* \text{ to } G_2 \quad \text{if } \beta_0 + \boldsymbol{\beta}'\mathbf{x} \leq -1.$$

The perpendicular distance between the two dotted lines is $2/||\boldsymbol{\beta}||$. The optimum β_0 and $\boldsymbol{\beta}$ are given by the solution of the quadratic programming problem

$$\text{minimize }_{\beta_0,\boldsymbol{\beta}} \tfrac{1}{2}||\boldsymbol{\beta}||^2$$
$$\text{subject to} \quad y_i(\beta_0 + \boldsymbol{\beta}'\mathbf{x}_i) \geq 1, \quad i = 1,\ldots,n.$$

Lagrange multipliers, α_i, are introduced for the inequality constraints, and these have to be positive. The quantity, L, now to be minimized is

$$L = \frac{1}{2}||\boldsymbol{\beta}||^2 - \sum_i \alpha_i y_i(\beta_0 + \boldsymbol{\beta}'\mathbf{x}_i) + \sum_i \alpha_i.$$

This is the *primal* optimization problem, of minimizing L with respect to β_0 and $\boldsymbol{\beta}$, such that $\partial L/\partial \alpha_i = 0$, and also that $\alpha_i \geq 0$. This is equivalent to the *dual* problem of maximizing L, such that $\partial L/\partial \beta_0 = 0$, and $\partial L/\partial \boldsymbol{\beta} = \mathbf{0}$, and subject to the constraint that $\alpha_i \geq 0$. Differentiating L,

$$\frac{\partial L}{\partial \beta_0} = -\sum_i \alpha_i y_i = 0$$

$$\frac{\partial L}{\partial \boldsymbol{\beta}} = \boldsymbol{\beta} - \sum_i \alpha_i y_i \mathbf{x}_i = \mathbf{0}.$$

From these equations, substituting $\boldsymbol{\beta} = \sum_i \alpha_i y_i \mathbf{x}_i$, and $\sum_i \alpha_i y_i = 0$ back into L, gives the dual problem as maximizing L, where

$$L = \sum_i \alpha_i - \frac{1}{2}\sum_i \sum_j \alpha_i \alpha_j y_i y_j \mathbf{x}_i'\mathbf{x}_j,$$

subject to the constraints $\sum_i \alpha_i y_i = 0$, $y_i(\beta_0 + \boldsymbol{\beta}'\mathbf{x}_i) - 1 = 0$, $\alpha_i(y_i(\beta_0 + \boldsymbol{\beta}'\mathbf{x}_i) - 1) = 0$ and $\alpha_i \geq 0$.

The α_i values that maximize L are denoted by $\hat{\alpha}_i$. It can be shown that $\hat{\alpha}_i = 0$ for all points except those for the support vectors. Once these have been obtained, the best fitting

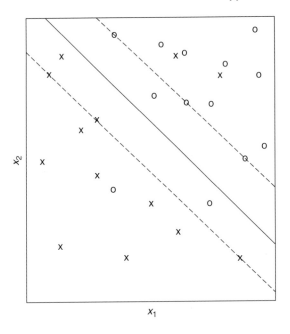

Figure 13.5 Overlap of the groups

discriminant has $\hat{\boldsymbol{\beta}} = \sum_i \hat{\alpha}_i y_i \mathbf{x}_i$, and β_0 estimated by $\hat{\beta}_0$, the mean of $y_i - \boldsymbol{\beta}' \mathbf{x}_i$, and with the mean being taken over the support vectors only. The discriminant is thus

$$h(\mathbf{x}) = \hat{\beta}_0 + \sum_i \hat{\alpha}_i y_i \mathbf{x}_i' \mathbf{x}.$$

In most practical situations, the two groups will not be separable, and so we allow some of the Group 1 points to cross over into the Group 2 region and vice versa, as shown in Figure 13.5. Slack variables, ξ_i, are introduced and the constraints become

$$\beta_0 + \boldsymbol{\beta}' \mathbf{x}_i \geq 1 - \xi_i \quad \text{for } y_i = 1$$
$$\beta_0 + \boldsymbol{\beta}' \mathbf{x}_i \leq -1 + \xi_i \quad \text{for } y_i = -1,$$

with the function to be minimized

$$L = \frac{1}{2}||\boldsymbol{\beta}||^2 + \alpha_0 \sum_i \xi_i - \sum_i \alpha_i \{y_i(\beta_0 + \boldsymbol{\beta}' \mathbf{x}_i) - (1 - \xi_i)\} - \sum_i \mu_i \xi_i.$$

The dual problem is again to minimize

$$L = \sum_i \alpha_i - \frac{1}{2} \sum_i \sum_j \alpha_i \alpha_j y_i y_j \mathbf{x}_i' \mathbf{x}_j,$$

subject to

$$0 \le \alpha_i \le \alpha_0,$$

$$\sum_i \alpha_i y_i = 0.$$

The estimate of $\boldsymbol{\beta}$ is again given by $\hat{\boldsymbol{\beta}} = \sum_i \hat{\alpha}_i y_i \mathbf{x}_i$, and the discriminant is $\hat{\beta}_0 + \sum_i \hat{\alpha}_i y_i \mathbf{x}_i' \mathbf{x}$. The discriminant is linear in $\mathbf{x}$, giving a linear discriminant boundary. We could transform the data to produce a better discriminant, by transforming $\mathbf{x}$ to $\mathbf{f}(\mathbf{x})$, where $\mathbf{f}(.)$ is a vector function taking $\mathbf{x}$ in a p-dimensional space, to $\mathbf{f}(\mathbf{x})$, in a q-dimensional space, where there is no restriction on q. The function now to be minimized is

$$L = \sum_i \alpha_i - \frac{1}{2} \sum_i \sum_j \alpha_i \alpha_j y_i y_j \langle \mathbf{f}(\mathbf{x}_i), \mathbf{f}(\mathbf{x}_j) \rangle,$$

where $\langle .,. \rangle$ denotes inner product. Write the inner product, $\langle \mathbf{f}(\mathbf{x}_i), \mathbf{f}(\mathbf{x}_j) \rangle$, as $K(\mathbf{x}_i, \mathbf{x}_j)$, known as the *kernel*, and so L becomes

$$L = \sum_i \alpha_i - \frac{1}{2} \sum_i \sum_j \alpha_i \alpha_j y_i y_j K(\mathbf{x}_i, \mathbf{x}_j).$$

The key advantage is that the actual transformation, $\mathbf{f}(.)$, does not have to be specified, only the kernel $K(.,.)$. Examples of kernels are

$$K(\mathbf{x}_i, \mathbf{x}_j) = (1 + \mathbf{x}_i' \mathbf{x}_j)^p$$

$$K(\mathbf{x}_i, \mathbf{x}_j) = \exp((\mathbf{x}_i - \mathbf{x}_j)'(\mathbf{x}_i - \mathbf{x}_j)).$$

Having chosen the kernel, L is minimized giving $\hat{\alpha}_i$ and the discriminant becomes

$$h(\mathbf{x}) = \hat{\beta}_0 + \sum_i \hat{\alpha}_i y_i K(\mathbf{x}_i, \mathbf{x}).$$

Again, we see that the advantage is that the actual transformation, $\mathbf{f}(\mathbf{x})$, does not have to be specified.

For readers interested in further details of SVMs, see Burges (1998) or Cristianini and Shawe-Taylor (2000) for example, and for software, see the website http://www.cs.cornell.edu/People/tj/svm_light/.

13.5 CART

One problem with discriminants that are a linear sum of variates, or possibly a more complicated function of them, is that they are not usually easily interpretable. It can be difficult to describe the discriminant to the layperson, who would rather have a discriminant that stated 'If the patient's systolic blood pressure is greater than 160, then assign the patient to Group 1; otherwise to Group 2.'

Classification and regression trees (CART) produce a series of discriminant rules of this nature. Breiman et al. (1984) describes them in detail. A tree is formed from our data

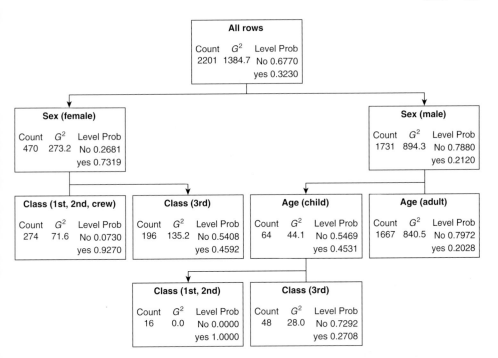

Figure 13.6 CART for *Titanic* survivors

matrix **X**. The tree consists of nodes that branch to other nodes, that in turn, branch into yet more nodes. Figure 13.6 shows a typical tree. Continuous, discrete, categorical and ordered categorical data can be analysed using CART. However, we will outline the technique for the case where we wish to classify observations into one of two groups, and will use the *Titanic* survival data described in Chapter 8 for illustration. CHAID (Chi-squared Automatic Interaction Detector) is an allied method for producing such trees.

The top node of the tree contains all the observations. This is split into two nodes according to a splitting rule used on one of the variables. The splitting rules are of the following types. In general, for continuous, discrete and ordered categorical variables:

Split the observations into two groups according to whether $x_i \leq c$ or not, where c is a constant.

For categorical variables:

If $c_1, \ldots, c_m$ are the categories, split the observations into two groups according to whether $x_i \in S$ or not, where S is a subset of $c_1, \ldots, c_m$.

The split is made that gives the optimum splitting of the observations in the node to form the two nodes below that node. A criterion for choosing which is the best split has to be chosen. One possibility is to define the *impurity* of a node as

$$-\sum_i p_i \log p_i,$$

where p_i is the proportion of observations in group i. This is based on a measure of entropy. Then the split is chosen that gives the largest decrease in impurity. So, for a particular split, if the two new nodes have proportions p_i^1 and p_i^2, for the proportions of observations in the groups, then the decrease in impurity is

$$\left(-\sum_i p_i \log p_i \right) - \left(-\sum_i p_i^1 \log p_i^1 \right) - \left(-\sum_i p_i^2 \log p_i^2 \right).$$

Another possible measure of impurity is the Gini diversity index, $\sum_{i \neq j} p_i p_j$. There are several other measures that can also be adopted.

The nodes of the tree are continually split into more and more nodes. However, at some stage, a decision has to be made to stop splitting. Otherwise the analyst could end up with a tree containing hundreds of nodes, the terminal nodes (the ones at the bottom of the branches) each containing one observation only. Trees can be *pruned* at a particular node by taking away all nodes emanating from that node, leaving the node as a terminal node. But how do we decide whether or not to prune at a node? One possibility is to assign each terminal node to one of the groups, according to the maximum of p_1 and p_2. Then the total number of misclassified observations within the terminal nodes is counted. As the tree grows this number will decrease, and as the tree is pruned, so will it increase. A very large tree can be constructed at first, and then pruned back, each pruning giving an increase in the total misclassified observations. Pruning should stop when there is a sharp increase in the total number of misclassified observations. We have presented a simple methodology here for illustration, and concentrated on classification trees. The same ideas apply for regression trees.

The statistical software package JMP was used to produce the tree in Figure 13.6 for the *Titanic* survival data. The three variables are categorical, and for these, JMP uses $G^2 = \sum_i p_i \log p_i$ for the impurity of a node. Looking at the tree, the first split is on *sex*, into females and males. Females are then split into {*1st Class, 2nd Class, Crew*}, and {*3rd Class*}. For the first group, 95% survived, but only 45% for the second. The male group split by *age* into children and adults. The male children are then split into {*1st Class, 2nd Class*}, and {*3rd Class*}, where 100% and 27% survived, respectively. The adult males are split into {*1st Class, Crew*}, and {*2nd Class, 3rd Class*}, where 24% and 14% survived, respectively.

13.6 Canonical variates analysis

All the methods so far discussed in this chapter have been designed to allocate an observation, of unknown origin, to one of two or more groups, and hence the term *discriminant analysis*. Canonical variates analysis is essentially the same as discriminant analysis, but with emphasis on the description of the differences between groups. What is it in the variables collected on patients that separates the group of prostate cancer patients from the non-cancer patients? Canonical variates are linear combinations of the x-variables, $y = a'x$, that best separate the groups. They are found as for Fisher's linear discriminant function, maximizing the ratio of the between-group variance of Y to the within-group variance of Y, as in (13.2). However, this time we do not simply use the largest eigenvalue of $W^{-1}B$, and its associated eigenvector a to define the linear discriminant, but use all the eigenvalues and their associated eigenvectors. These define the set of *canonical variates*, ordered according to size of the eigenvalues. The total number of canonical variates that can be obtained is $\min(p, G - 1)$, where G is the number of groups.

Canonical variates are not orthogonal, i.e. $\mathbf{a}_i'\mathbf{a}_j \neq 0$, and are thus correlated. They are not a rigid rotation of the axes, as are principal components. However, within groups, the canonical variates are uncorrelated. They are also not scale dependent. They are usually scaled to have unit within-group variance, i.e. $\mathbf{a}'\mathbf{Wa} = 1$. This is done to aid in the interpretation of the coefficients of $\mathbf{a}$, since individual coefficients may be large because, either the corresponding x-variable is good at separating the groups, or has small within-group variance.

Example

The three canonical variates for the polychaete data are given in Table 13.1

For each specimen the canonical variate scores can be calculated. Figure 13.7 shows the scores for the second canonical variate plotted against those for the first. The configuration of points forms an interesting horseshoe shape, with points for species 1 and 2 intermingled at the centre, points for species 3 to the left and points for species 4 to the right. The first canonical variate is very good at separating the four species on its own. The second canonical variate helps separate the species even further.

Table 13.1

Variable	Can 1	Can 2	Can 3
area	−3.74	−4.87	−2.60
perimeter	−2.92	−0.78	−6.00
elongation	1.24	1.05	0.61
roundness	−0.57	0.14	−1.17
feret	9.64	5.40	8.98
Eigenvalue	6.263	0.497	0.133

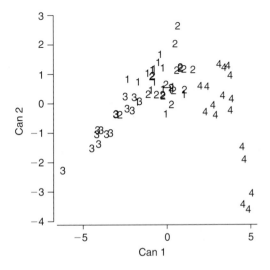

Figure 13.7 Canonical variate scores for the four species of polychaete

13.7 Exercises

1. The training data for two groups gave rise to the following mean vectors and pooled sample covariance matrix

$$\bar{x}_2 = \begin{bmatrix} 2.0 \\ 2.3 \end{bmatrix} \quad S_u = \begin{bmatrix} 3.2 & 1.3 \\ 1.3 & 4.8 \end{bmatrix}.$$

Assuming a multivariate normal distribution, find the (estimated) maximum likelihood discriminant rule. Draw the discriminant boundary on a plot of x_2 against x_1.
   Suppose there had been a third group, that had mean vector

$$\bar{x}_1 = \begin{bmatrix} 3.0 \\ 0.9 \end{bmatrix}.$$

Using the same pooled covariance matrix as above, find the discriminant boundaries between Group 1 and Group 3, and between Group 2 and Group 3. Draw these on the plot.

2. Carry out linear discrimination on the prostate cancer data described in Chapter 4, using svi – seminal vesicle invasion, to define two groups, and lcp – log capsular penetration, and $lpsa$ – log prostate specific antigen as the x-variables. Try incorporating other variables into the discrimination.

3. Let the probability that an observation belongs to Group 1, given $\mathbf{x} = (x_1, x_2)'$, be

$$p_1(\mathbf{x}) = \frac{e^{\beta_0 + \beta_1 x_1 + \beta_2 x_2}}{1 + e^{\beta_0 + \beta_1 x_1 + \beta_2 x_2}},$$

and the probability that it belongs to Group 2, $p_2(\mathbf{x}) = 1 - p_1(\mathbf{x})$. The training data have n_i observations for Group i, $(i = 1, 2)$.
   Write down the likelihood and then find the log-likelihood.
   Find the likelihood equations for estimating β_0, β_1 and β_2.

4. Carry out logistic discrimination on the prostate cancer, again using svi – seminal vesicle invasion, as the group membership variable, and lcp – log capsular penetration, and $lpsa$ – log prostate specific antigen as the x-variables.

5. Fisher's iris data is a famous dataset that can be found at several websites, for example the DASL website. The data consist of the four measurements, *sepal length, sepal width, petal length* and *petal width,* for three species of iris, *Iris setosa, Iris versicolour* and *Iris virginica.* These measurements are made on fifty flowers of each species. Download the data and carry out a canonical variates analysis.

Loglinear modelling

This chapter introduces the reader to loglinear models, firstly for modelling the mean of the Poisson distribution, and then for modelling contingency tables. Only a snapshot can be given here, and for more detailed accounts see Agresti (2002) and McCullagh and Nelder (1989).

14.1 Loglinear models for Poisson means

We start with an example where we model the number of goals scored in football matches by the team playing at home. Table 14.1 gives the frequency count of the number of home goals in a sample of fifty matches played in the UK, one Saturday afternoon, during 2004.

Table 14.1 Frequency of the number of home goals

No. of home goals	0	1	2	3	4	5	6	Total
Frequency	9	22	9	7	2	0	1	50

Suppose the number of goals scored by the home side, Y, follows a Poisson distribution with mean μ. So, $\Pr(Y = y) = \mu^y e^{-\mu}/y!$, ($y = 0, 1, 2, \ldots$). The likelihood is

$$L = \prod_{i=1}^{n} \left(\frac{\mu^{y_i} e^{-\mu}}{y_i!} \right). \tag{14.1}$$

The log-likelihood is

$$l = \log L = \left(\sum_{i=1}^{n} y_i \right) \log \mu - n\mu - \sum_{i=1}^{n} \log y_i!.$$

We now make the transformation

$$\log \mu = \beta_0, \tag{14.2}$$

and hence

$$l = \left(\sum_{i=1}^{n} y_i \right) \beta_0 - n e^{\beta_0} - \sum_{i=1}^{n} \log y_i!.$$

Maximizing the log-likelihood gives the maximum likelihood estimate of β_0 as

$$\hat{\beta}_0 = \log\left\{n^{-1}\sum_{i=1}^{n} y_i\right\} = \log \bar{y}.$$

Now this is standard maximum likelihood estimation for the mean of the Poisson distribution, but has been introduced in this way, with the transformation (14.2), to illustrate the idea of loglinear models. For a loglinear model, the log of the mean is linear in the parameters (here we only have one parameter, namely β_0). For our data, $\bar{y} = 1.50$, and so $\hat{\beta}_0 = 0.4055$.

Now suppose we have one or more explanatory variables, $x_1, \ldots, x_p$. These can be incorporated into the model thus

$$\log \mu = \beta_0 + \beta_1 x_1 + \beta_2 x_2 + \ldots + \beta_p x_p,$$

and then the maximum likelihood estimates, $\hat{\beta}_0, \hat{\beta}_1, \ldots, \hat{\beta}_p$ can be found. Fitted values, $\hat{y}_i$, are given by

$$\hat{y}_i = \exp(\hat{\beta}_0 + \hat{\beta}_1 x_{1i} + \ldots + \hat{\beta}_p x_{pi}),$$

where $x_{1i}, \ldots, x_{pi}$ are the values of the explanatory variables for the ith data point.

A measure of how well the model fits the data is given by the *deviance*, D, where

$$D = 2(l(\mathbf{y}; \mathbf{y}) - l(\mathbf{y}; \boldsymbol{\beta})),$$

where $l(\mathbf{y}; \mathbf{y})$ is the value of the log-likelihood when there are as many parameters as there are data points, and $l(\mathbf{y}; \hat{\boldsymbol{\beta}})$ is the value of the log-likelihood under the model. The value $l(\mathbf{y}; \mathbf{y})$ is the maximum achievable value of the log-likelihood; it is the *saturated* model where $\hat{y}_i = y_i$, i.e. the fitted values equal the actual data values. The deviance is not just for the Poisson distribution, but for all generalized linear models (see McCullagh and Nelder, 1989, or Dobson, 2001). Asymptotically, D has a chi-squared distribution, with degrees of freedom equal to the difference in the number of parameters between the saturated model and the particular model being fitted. Replacing μ by μ_i in (14.1), the log-likelihood for the saturated model is

$$l = \sum_{i=1}^{n} y_i \log \mu_i - \sum_{i=1}^{n} \mu_i - \sum_{i=1}^{n} \log y_i!.$$

Maximizing the log-likelihood gives $\hat{\mu}_i = y_i$, and thus

$$l(\mathbf{y}; \mathbf{y}) = \sum_{i=1}^{n} y_i \log y_i - \sum_{i=1}^{n} y_i - \sum_{i=1}^{n} \log y_i!.$$

The log-likelihood for the model is given by

$$l(\mathbf{y}; \boldsymbol{\beta}) = \sum_{i=1}^{n} y_i \log \hat{y}_i - \sum_{i=1}^{n} \hat{y}_i - \sum_{i=1}^{n} \log y_i!.$$

Hence, the deviance is

$$2\sum_{i=1}^{n}\{y_i \log(y_i/\hat{y}_i) - (y_i - \hat{y}_i)\}.$$

Returning to the football data, we will use the current position in the league as an explanatory variable, x_1, for the number of goals scored by the home team. We will use this as a continuous variable rather than an ordered categorical variable. The estimates, $\hat{\beta}_0$ and $\hat{\beta}_1$, of β_0 and β_1, are 0.5786 and -0.0162, with estimated standard errors of 0.2179 and 0.0179, respectively. The deviance for the model is 53.12 on 48 degrees of freedom. From the chi-squared distribution, this has a p-value of 0.283, and so the model appears to fit the data. When the position in the league is not included in the model, the deviance is 53.96 on 49 degrees of freedom. We can now test whether the position in the league is actually needed in the model. This is done by looking at the difference in the deviances. The difference will asymptotically have a chi-squared distribution with degrees of freedom equal to the difference of the degrees of freedom for the two models. The difference in the deviances is 0.84. From the chi-squared distribution with one degree of freedom, this has a p-value of 0.359, and hence we can conclude that position in the league does not help us to predict the number of goals scored by the home team.

14.2 Contingency tables

We now look at loglinear models for contingency tables, again using football scores as an example. Suppose we now record, not only the score by the home team, but also the score by the away team. Let p_{ij} be the probability that the home team scores $i - 1$ goals and the away team $j - 1$ goals. However, we will now amalgamate all scores greater than or equal to four, to one category labelled '≥ 4'. Table 14.2 shows the joint probability distribution for (home score, away score). The *dot notation* is used for the row sums of the probabilities, and thus $p_{i.} = \sum_{j=1}^{5} p_{ij}$, and similarly for the column sums, $p_{.j} = \sum_{i=1}^{5} p_{ij}$. The row sums, $p_{1.}, \ldots, p_{5.}$ give the marginal distribution of the home score, and the column sums, $p_{.1}, \ldots, p_{.5}$, the marginal distribution of the away score. The observed data are summarized in a contingency table shown in Table 14.3.

We must think about how the data were actually collected. The total number of matches observed was 50, and we condition on this number being fixed. Even if the total number of matches was not fixed, we would usually conduct our statistical inference based upon it being

Table 14.2 Probability distribution for (home score, away score)

		Away score					
		0	1	2	3	≥ 4	
Home score	0	p_{11}	p_{12}	p_{13}	p_{14}	p_{15}	$p_{1.}$
	1	p_{21}	p_{22}	p_{23}	p_{24}	p_{25}	$p_{2.}$
	2	p_{31}	p_{32}	p_{33}	p_{34}	p_{35}	$p_{3.}$
	3	p_{41}	p_{42}	p_{43}	p_{44}	p_{45}	$p_{4.}$
	≥ 4	p_{51}	p_{52}	p_{53}	p_{54}	p_{55}	$p_{5.}$
		$p_{.1}$	$p_{.2}$	$p_{.3}$	$p_{.4}$	$p_{.5}$	$p_{..}$

Table 14.3 Contingency table (home score, away score)

		Away score					
		0	1	2	3	≥ 4	
Home score	0	n_{11}	n_{12}	n_{13}	n_{14}	n_{15}	$n_{1.}$
	1	n_{21}	n_{22}	n_{23}	n_{24}	n_{25}	$n_{2.}$
	2	n_{31}	n_{32}	n_{33}	n_{34}	n_{35}	$n_{3.}$
	3	n_{41}	n_{42}	n_{43}	n_{44}	n_{45}	$n_{4.}$
	≥ 4	n_{51}	n_{52}	n_{53}	n_{54}	n_{55}	$n_{5.}$
		$n_{.1}$	$n_{.2}$	$n_{.3}$	$n_{.4}$	$n_{.5}$	$n_{..}(=n)$

fixed. The distribution of the frequencies $n_{11}, n_{12}, \ldots, n_{21}, \ldots, n_{rc}$, follows a multinomial distribution, where now r denotes the number of rows, and c the number of columns. The probabilities are given by

$$\Pr(n_{11}, \ldots, n_{rc}) = \frac{n!}{n_{11}! \ldots n_{rc}!} p_{11}^{n_{11}} \ldots p_{rc}^{n_{rc}}. \tag{14.3}$$

The log-likelihood is simply the log of this expression,

$$l = \log\left(\frac{n!}{n_{11}! \ldots n_{rc}!}\right) + \sum_{i=1}^{r} \sum_{j=1}^{c} n_{ij} \log p_{ij}.$$

The mean of n_{ij} is np_{ij} ($1 \leq i \leq r; 1 \leq j \leq c$). The loglinear model for the means is

$$\log(np_{ij}) = \theta + \alpha_i + \beta_j + \alpha\beta_{ij},$$

where α represents a row effect, β a column effect, and $\alpha\beta$ a row–column interaction. (Note that $\alpha\beta$ is a parameter and not a product of parameters.) For example, the mean of n_{23}, from the second row, third column, is modelled as

$$\log(np_{23}) = \theta + \alpha_2 + \beta_3 + \alpha\beta_{23},$$

or equivalently

$$np_{23} = e^{\theta + \alpha_2 + \beta_3 + \alpha\beta_{23}}.$$

We have rc n's, whilst the model has 1 θ, r α's, c β's and rc $\alpha\beta$'s, giving a total of $1 + r + c + rc$ parameters. The model is thus over-parameterized, and we have to reduce the number of parameters in some way. There are two common ways of doing this. The first is to set some of the parameters to zero. We set any parameter that has a '1' in its suffices equal to zero. Thus $\alpha_1 = \beta_1 = \alpha\beta_{11} = \ldots = \alpha\beta_{1c} = \alpha\beta_{21} = \ldots = \alpha\beta_{r1} = 0$, which is a total of $r + c + 1$ zeros, and hence the number of free paramters is now rc. The number of parameters equals the number of data points – we have a saturated model that will fit perfectly. However, in practice, there is little point in modelling rc data points with rc parameters. The other way to reduce the number of parameters is to make the following restrictions:

$$\sum_{i=1}^{r} \alpha_i = 0, \qquad \sum_{j=1}^{c} \beta_j = 0,$$

$$\sum_{i=1}^{r} \alpha\beta_{ij} = 0, (j = 1, \ldots, c), \qquad \sum_{j=1}^{c} \alpha\beta_{ij} = 0, (i = 1, \ldots, r),$$

again reducing the number of free parameters down to rc. There is actually one more restriction in the fact that we are conditioning on the total number, n, being fixed.

The hypothesis that we are interested in is whether the marginal distribution of the rows is independent of the marginal distribution of the columns. For us, is the home score independent of the away score? If so, then

$$p_{ij} = p_{i.}p_{.j}. \tag{14.4}$$

If $\alpha\beta_{ij} = 0$, $(i = 1, \ldots, r; j = 1, \ldots, c)$, in the saturated model, then

$$p_{ij} = n^{-1}e^{\theta+\alpha_i+\beta_j}, \tag{14.5}$$

and hence

$$p_{i.} = n^{-1}e^{\theta+\alpha_i}\sum_{j=1}^{c}e^{\beta_j}$$

$$p_{.j} = n^{-1}e^{\theta+\beta_j}\sum_{i=1}^{r}e^{\alpha_i}$$

$$p_{..} = n^{-1}e^{\theta}\sum_{i=1}^{r}\alpha_i\sum_{j=1}^{c}e^{\beta_j}.$$

Now $p_{..} = 1$ and thus

$$1 = n^{-1}e^{\theta}\sum_{i=1}^{r}\alpha_i\sum_{j=1}^{c}e^{\beta_j}. \tag{14.6}$$

Multiplying equation (14.5) by (14.6)

$$p_{ij} = n^{-1}e^{\theta+\alpha_i+\beta_j} \times n^{-1}e^{\theta}\sum_{i=1}^{r}e^{\alpha_i}\sum_{j=1}^{c}e^{\beta_j}$$

$$= n^{-1}e^{\theta+\alpha_i}\sum_{j=1}^{c}e^{\beta_j} \times n^{-1}e^{\theta+\beta_j}\sum_{i=1}^{r}e^{\alpha_i}$$

$$= p_{i.}p_{.j}.$$

Thus $\alpha\beta_{ij} = 0$ $(i = 1, \ldots, r; j = 1, \ldots, c)$ in the saturated model corresponds to independent marginal distributions.

Table 14.4 shows the contingency table for the fifty football matches. Table 14.5 gives the parameter estimates together with their estimated standard errors for the independent marginal distribution model.

The deviance is 23.61 with 16 degrees of freedom, which has a p-value of 0.098. This p-value is used to assess the fit of the model, and also corresponds to that for testing the hypothesis of independent marginal distributions. If, because of the low frequency counts, the categories '3' and '≥ 4' are amalgamated for both home and away scores, and the model refitted, the deviance becomes 16.21 on nine degrees of freedom, which has a p-value of 0.063. Thus the evidence, although not very strong, points to us rejecting the null hypothesis

Table 14.4 Football contingency table

| | | \multicolumn{5}{c}{Away score} | |
		0	1	2	3	≥ 4	
	0	4	1	2	1	1	9
	1	9	8	4	1	0	22
Home score	2	0	5	2	2	0	9
	3	3	1	0	2	1	7
	≥ 4	1	2	0	0	0	3
		17	17	8	6	2	50

Table 14.5 Estimates of parameters

Parameter	Estimate	St. error	χ^2	p
θ	1.12	0.387	8.34	0.004
α_1	0	0	.	.
α_2	0.89	0.40	5.10	0.024
α_3	0.00	0.47	0.00	1.000
α_4	−0.25	0.50	0.25	0.618
α_5	−1.10	0.67	2.72	0.099
β_1	0	0	.	.
β_2	0.00	0.34	0.00	1.000
β_3	−0.75	0.43	3.09	0.079
β_4	−1.04	0.47	4.81	0.028
β_5	−2.14	0.75	8.20	0.004

of independent marginal distributions. However, those who stick to the rule of rejecting the null hypothesis only when the p-value is less than 0.05 would declare that there was not enough evidence to reject the hypothesis of independent marginal distributions.

14.3 Higher dimensional tables

The loglinear model can be extended to higher dimensions. But before doing this, let us consider how the football data in the previous section were collected, and possible alternatives. For the football data, fifty matches were selected overall, and then the home scores and away scores were noted. There are two response variables, home score and away score. The total number of matches was fixed at fifty. However, the row totals and column totals are random. Thus the only constraint is that cell frequencies add to n, i.e.

$$\sum_{i=1}^{r} \sum_{j=1}^{c} n_{ij} = n.$$

The appropriate distribution is the multinomial distribution (14.2). This design is referred to as a *cross-sectional study* as it takes data at a snapshot in time, one Saturday afternoon.

Changing subject, suppose we wish to investigate the efficacy of three drugs (A, B and C) for curing a particular disease and wish to find the best. We would set up a clinical trial

Table 14.6 Contingency table for drug comparison

		Cure			
		No cure	Partial cure	Cure	
	A	n_{11}	n_{12}	n_{13}	$n_{1.}$
Drug	B	n_{21}	n_{22}	n_{23}	$n_{2.}$
	C	n_{31}	n_{32}	n_{33}	$n_{3.}$
		$n_{.1}$	$n_{.2}$	$n_{.3}$	$n_{..}$

where each drug was tested by one hundred patients, say. Suppose the response to a drug is: not cured, partially cured, cured. After 6 months from starting the trial, the responses are collected and a contingency table is formed as in Table 14.6. This type of study is a *prospective study*, as we look forward in time.

There is only one response variable, 'cured'. The other variable, 'drug', is an explanatory variable. The row sums are all fixed to be equal to one hundred, and this has to be reflected in any model for the data. The appropriate distribution for the cell frequencies is the *product multinomial*

$$\Pr(n_{11}, \ldots, n_{rc}) = \prod_{i=1}^{r} \left\{ \frac{n_{i.}!}{n_{i1}! \ldots n_{ic}!} p_{i1}^{n_{i1}} \ldots p_{ic}^{n_{ic}} \right\},$$

which is simply the product of multinomial distributions, one for each row.

The prospective study described above is not always feasible, for instance, if the time that has to elapse before obtaining the data is excessive, or if the study would not be allowed on ethical grounds. For instance, a study where one group of subjects were forced to smoke sixty cigarettes a day for five years, to investigate the effects of smoking on health, would not be passed by an ethics committee. Instead, a *retrospective study* is needed, where the smoking habits over the lifetime of a group of patients with cancer, and also for a matched group of healthy people, are studied in order to shed light on what might be increasing the risk of cancer. Suppose smoking habits are classified as heavy smoker/light smoker/non-smoker. The response variable is 'smoking', and the explanatory variable is 'cancer'.

The reason for considering the above sampling scenarios is to show that any model for the data must take into account the particular row and column restrictions required by the sampling scheme.

The three-dimensional loglinear model is

$$\log(np_{ijk}) = \theta + \alpha_i + \beta_j + \gamma_k + \alpha\beta_{ij} + \alpha\gamma_{ik} + \beta\gamma_{jk} + \alpha\beta\gamma_{ijk}.$$

As for the two-dimensional model, the model is over-parameterized and so restrictions are placed on the parameters. For instance we set

$$\alpha_1 = \beta_1 = \gamma_1 = 0$$

$$\alpha\beta_{1j} = \alpha\beta_{i1} = \alpha\gamma_{1k} = \alpha\gamma_{i1} = \beta\gamma_{1k} = \beta\gamma_{j1} = 0$$

$$\alpha\beta\gamma_{1jk} = \alpha\beta\gamma_{i1k} = \alpha\beta\gamma_{ij1} = 0.$$

Example

The data in Table 14.7 concern lung cancer in mice that were exposed to Avadex (Innes et al., 1969). There are three explanatory variables: strain, sex and exposure. The response variable is whether a tumour develops or not.

For a quick exploratory analysis of the data, Table 14.8 shows the ratio of the number of mice with a tumour to the number without a tumour for each of the groups, defined by the rows of Table 14.8. It appears that exposure to Avadex lessens the risk of a tumour developing. Other questions of interest are whether there is a difference in risk between the two strains of mice, and whether there is a difference between male and female. We use loglinear models to explore the data further.

We will consider only strain X mice first. Now, the rows of the table are fixed as the experiment had 16 male mice of strain X exposed to Avadex, 79 male mice of strain X used as a control, and so on. These fixed quantities must be reflected in the model. To do this sex, exposure and the interaction sex*exposure must be included in the model. If these are the only terms used in the model, then the deviance is 178.9 with four degrees of freedom. The estimated parameters (including the ones that are set to zero) are shown in Table 14.9.

The estimated cell frequency for male exposed mice with a tumour is

$$n\hat{p}_{111} = \exp(\hat{\theta} + \hat{\alpha}_1 + \hat{\beta}_1 + \widehat{\alpha\beta}_{11})$$
$$= \exp(2.0794) = 8.0.$$

Similarly, the estimated cell frequency for male exposed mice without a tumour is also

$$n\hat{p}_{112} = \exp(\hat{\theta} + \hat{\alpha}_1 + \hat{\beta}_1 + \widehat{\alpha\beta}_{11}) = 8.0,$$

Table 14.7 Lung cancer in mice

Strain	Sex	Exposure	Tumour	No tumour
X	M	Exposed	12	4
		Control	74	5
X	F	Exposed	12	2
		Control	84	3
Y	M	Exposed	14	4
		Control	80	10
Y	F	Exposed	14	1
		Control	79	3

Table 14.8 Ratio of occurrence to non-occurrence

Strain	Sex	Exposure	Ratio
X	M	Exposed	3.0
		Control	14.8
X	F	Exposed	6.0
		Control	28.0
Y	M	Exposed	3.5
		Control	8.0
Y	F	Exposed	14.0
		Control	26.3

since tumour has not been included in the model. Note that the sum of the two estimates is sixteen, the correct value for the row sum. A similar pattern emerges for the other three rows in the table for the X strain. Each estimated cell frequency is half the row sum.

We now include the term for the tumour. If all the second-order interaction terms and the third-order interaction term are included in the model then the deviance is zero with a perfect fit of the model. The estimated parameters, excluding the ones set to zero, are shown in Table 14.10. The reader is invited to check that the estimated cell frequencies are identical to the observed frequencies.

Removing various terms involving tumour from the saturated model and fitting the subsequent models gave the deviances shown in Table 14.11.

Generally, if the model is chosen where all three-factor and two-factor interactions are absent, apart from those necessary to guarantee row sums, then there is independence between the variables. If the three-factor interaction and some of the two-factor interactions are absent, then conditional independence occurs. From the table of deviances below,

Table 14.9 Parameter estimates

Parameter		Estimate	St. error
Intercept	θ	2.0794	0.2500
sex M	α_1	0	0
sex F	α_2	−0.1335	0.3660
exp E	β_1	0	0
exp C	β_2	1.5969	0.2742
sex*exp (M*E)	$\alpha\beta_{11}$	0	0
sex*exp (F*E)	$\alpha\beta_{21}$	0	0
sex*exp (M*C)	$\alpha\beta_{12}$	0	0
sex*exp (F*C)	$\alpha\beta_{22}$	0.2300	0.3976

Table 14.10 Parameter estimates

Parameter		Estimate
Intercept	θ	2.4849
sex F	α_2	0.0000
exp C	β_2	1.8192
tum N	γ_2	−1.0986
sex*exp (F*C)	$\alpha\beta_{22}$	0.1268
sex*tum (F*N)	$\alpha\gamma_{22}$	−0.6931
exp*tum (C*N)	$\beta\gamma_{22}$	−1.5960
sex*exp*tum (F*C*N)	$\alpha\beta\gamma_{222}$	0.0556

Table 14.11 Deviance table

Model		Deviance	df
sex+exp+sex*exp		178.4	4
sex+exp+sex*exp	+tum	8.0151	3
sex+exp+sex*exp	+tum+sex*tum	6.4901	2
sex+exp+sex*exp	+tum+exp*tum	1.2992	2
sex+exp+sex*exp	+tum+sex*tum+exp*tum	0.0021	1
sex+exp+sex*exp	+tum+sex*tum+exp*tum+sex*exp*tum	0.0	0

and comparing the values with the upper 5% points of the chi-squared distribution, the parsimonious model which fits the data best is

$$(sex+exp+sex * exp)+tum+exp * tum.$$

This model tells us that whether a tumour occurs is related to whether there was exposure to Avadex. The difference in deviances for this model and the model with both two-factor interactions is 1.271 on 1 df. This reduction in deviance is not significant and hence we prefer to keep the model with only the one two-factor interaction.

We now extend the model to include the strain of mice. This time strain, sex and exposure and all their interactions have to be included in the model to ensure the correct row sums. The deviances for the various models are given in Table 14.12, where ( ) indicates the terms needed to ensure the correct row sums.

Although several of the models fit well, the parsimonious one we choose is

$$( )+tum+sex * tum+exp * tum$$

which again tells us that exposure to Avadex affects whether a tumour occurs, and also that the sex of the mouse affects whether a tumour occurs. Note that interactions with strain do not occur and so the tumour rates are independent of strain.

Table 14.12 Deviance table

Model	Deviance	df
()	348.1	8
()+t	15.38	7
()+t+st*t	15.02	6
()+t+se*t	8.89	6
()+t+ex*t	7.92	6
()+t+st*t+se*t	8.63	5
()+t+st*t+ex*t	7.58	5
()+t+se*t+ex*t	1.72	5
()+t+st*t+se*t+ex*t	1.47	4
()+t+st*t+se*t+ex*t+st*se*t	0.97	3
()+t+st*t+se*t+ex*t+st*ex*t	0.52	3
()+t+st*t+se*t+ex*t+se*ex*t	1.47	3
()+t+st*t+se*t+ex*t+st*se*t+st*ex*t	0.03	2
()+t+st*t+se*t+ex*t+st*se*t+se*ex*t	0.97	2
()+t+st*t+se*t+ex*t+st*ex*t+se*ex*t	0.52	2
()+t+st*t+se*t+ex*t+st*se*t+st*ex*t+se*ex*t	0.006	1

(t – tumour, st – strain, se – sex, ex – exposure)

14.4 Exercises

1. Collect the scores from fifty football matches, together with the position of the teams in their league. Model the number of home goals scored with a Poisson distribution, and use the difference between the league positions of the home and away teams as a covariate. Do the same for the number of goals scored by the away teams.

2. Consider a set of random variables, $\{Y_i\}$, $i=1,\ldots,I$, independently distributed, each with a Poisson distribution with individual parameter, λ_i. The joint distribution

of the Y's is

$$f(\mathbf{y}; \boldsymbol{\lambda}) = \prod_{i=1}^{I} \frac{\lambda_i^{y_i} e^{-\lambda_i}}{y_i!}.$$

Show that the distribution of $N = \sum_{i=1}^{I} Y_i$ also has a Poisson distribution.
   Show that the distribution of the Y's, conditional on N has a multinomial distribution.
How does this result relate to the analysis of contingency tables?

3. Consider the 2×2 contingency table

	A		Total
B	n_{11}	n_{12}	$n_{1.}$
	n_{21}	n_{22}	$n_{2.}$
Total	$n_{.1}$	$n_{.2}$	$n_{..}$

The saturated loglinear model for the data is

		A		Total	
B	e^{θ}	$e^{\theta+\alpha}$		$e^{\theta} + e^{\theta+\alpha}$	
	$e^{\theta+\beta}$	$e^{\theta+\alpha+\beta+\alpha\beta}$		$e^{\theta+\beta} + e^{\theta+\alpha+\beta+\alpha\beta}$	
Total	$e^{\theta} + e^{\theta+\beta}$	$e^{\theta+\alpha} + e^{\theta+\alpha+\beta+\alpha\beta}$		$e^{\theta} + e^{\theta+\alpha} + e^{\theta+\beta} + e^{\theta+\alpha+\beta+\alpha\beta}$	

Show that the parameter estimates are given by

$$\hat{\theta} = \log n_{11}$$
$$\hat{\alpha} = \log(n_{12}/n_{11})$$
$$\hat{\beta} = \log(n_{21}/n_{11})$$
$$\widehat{\alpha\beta} = \log(n_{11}n_{22}/n_{12}n_{21}).$$

Show that the likelihood for the independence model ($\alpha\beta = 0$), ignoring the factor involving the $n_{ij}!$'s, is

$$\left(\frac{e^{\theta}}{C}\right)^{n_{11}} \left(\frac{e^{\theta+\alpha}}{C}\right)^{n_{12}} \left(\frac{e^{\theta+\beta}}{C}\right)^{n_{21}} \left(\frac{e^{\theta+\alpha+\beta}}{C}\right)^{n_{22}},$$

where

$$C = e^{\theta} + e^{\theta+\alpha} + e^{\theta+\beta} + e^{\theta+\alpha+\beta}.$$

Find the log-likelihood and hence show that the estimates of the parameters are given by

$$\hat{\theta} = \log(n_{.2}/n_{.1})$$
$$\hat{\alpha} = \log(n_{2.}/n_{1.})$$
$$\hat{\beta} = \log(n_{.1}n_{1.}/n_{..}).$$

Now find the deviance for the model.

A new test for a disease was carried out on forty patients. The following table is a 2×2 contingency table of those patients testing positive or negative for the disease, by the subsequent diagnosis as to whether the patients had the disease or not.

	Test negative	Test positive	Total
No disease	15	8	23
Disease	6	11	17
Total	21	19	40

Test the independence of the test results and the actual diagnosis of the disease. Compare the results with the usual Pearson's chi-squared test and Fisher's exact test for contingency tables.

4. Use loglinear models to analyse the malignant melanoma data in Exercise 1 of Chapter 7.

15

Factor analysis

Factor analysis attempts to find latent variables (hidden variables) which cannot be observed, that 'drive' the variables that can be observed, called *manifest variables*. A classic example is the measurement of intelligence. We do not have an instrument that we can scan over a subject's brain and obtain a reading of intelligence. Intelligence is a latent variable. In order to measure it, we might set our subject a series of tests, and from the results, obtain a measure of intelligence, IQ for instance. The test results are the observed variables. What do we mean by intelligence? Logical thinking? A good memory? Mathematical ability? Linguistic ability? Instead of a single measure of intelligence, we might wish to have several measures, and devise tests to measure these. Factor analysis does not actually start with preconceived ideas of the latent variables, but attempts to find them from the data.

Let the manifest variables be $X_1, \ldots, X_p$, with joint density function (or probability function) $f(\mathbf{x})$. Let the latent variables be $Y_1, \ldots, Y_m$, with joint density function, $h(\mathbf{y})$. Let $g(\mathbf{x}|\mathbf{y})$ be the conditional density function of $\mathbf{x}$ given $\mathbf{y}$. Then

$$f(\mathbf{x}) = \int h(\mathbf{y}) g(\mathbf{x}|\mathbf{y}) \, d\mathbf{y}.$$

The aim of factor analysis and other latent variable models is to find the Y's that make the X's conditionally independent, and if this is the case, then

$$g(\mathbf{x}|\mathbf{y}) = \prod_{i=1}^{p} g(x_i|\mathbf{y}),$$

and hence

$$f(\mathbf{x}) = \int h(\mathbf{y}) \prod_{i=1}^{p} g(x_i|\mathbf{y}) \, d\mathbf{y}. \tag{15.1}$$

This chapter covers factor analysis, which is the most commonly used latent variable method, while Chapter 16 covers other models.

There are two main methods for factor analysis, one based on the rotation of the principal components from a principal components analysis, and one based on a modelling approach using maximum likelihood. The first method sometimes causes confusion, with principal components analysis being confused with factor analysis, and vice versa. A detailed account of factor analysis can be found in Bartholomew and Knott (1999).

15.1 Principal components factor analysis

We start with the principal components method for factor analysis. With principal components factor analysis, first q principal components (PC's) are found, and then these are rotated so that they line up more with some of the original variables. An argument is that often many coefficients in a PC tend to have similar values, and interpretation of the PC is then difficult. If the coefficients were such that only a few were substantial in value, the rest being approximately zero, then interpretation would be much easier. However, in doing this, the role of the PCs is now changed, from a linear combination of the original variables which explain as much variation as possible, to one of latent variables.

We illustrate the technique with an example based on a consumer study about hand creams.

Example

A consumer study was carried out, where panellists tested, and commented upon three hand creams. Each hand cream was tested by fifty panellists. Various questions were asked, and the panellists had to score the product they were testing on a scale 1–5, indicating the following

1 – weak, 2 – slightly weak, 3 – average, 4 – slightly strong, 5 – very strong,

or other similar connotations, depending on the question. Although these are essentially categorical data, we treat the responses as numerical for this analysis, using scores $1, \ldots, 5$. Note that we do not have to use these scores just because the categories are labelled 1–5. For instance, we may wish to boost the 'very strong' category by giving it a score of ten, say.

The questions asked of the panellist, in abbreviated form, were:

Q1 – overall opinion	*Q11 – applied right amount*
Q2 – smooth application	*Q12 – sticky whilst applying*
Q3 – dried quickly	*Q13 – greasy whilst applying*
Q4 – pleasant fragrance	*Q14 – sticky after applying*
Q5 – protected hands	*Q15 – speed of drying*
Q6 – kept hands fresh	*Q16 – sticky during the day*
Q7 – wait till dry	*Q17 – greasy during the day*
Q8 – feel greasy	*Q18 – overall protection to hands*
Q9 – applied easily	*Q19 – fragrance*
Q10 – dosed from pack	*Q20 – fragrance upon application*

A principal components analysis was carried out using the correlation matrix for questions Q2–Q20, pooled over all three products. Table 15.1 shows the eigenvalues and eigenvectors giving the first five PC's. The PC's are

$$PC1 = 0.50 \times Q2 + 0.78 \times Q3 + \cdots - 0.03 \times Q20$$

$$\vdots \qquad \vdots \qquad \vdots \qquad \vdots$$

$$PC5 = 0.02 \times Q2 - 0.21 \times Q3 + \cdots - 0.14 \times Q20.$$

Table 15.1 Principal components of the consumer data

	PC1	PC2	PC3	PC4	PC5
Eigenvalue	5.75	2.38	2.20	1.74	1.27
Prop. of var.	30%	13%	12%	9%	7%
Q2	0.50	0.12	0.51	0.11	0.02
Q3	0.78	−0.20	−0.22	−0.29	−0.21
Q4	0.29	0.44	0.50	−0.49	0.21
Q5	0.26	0.70	−0.46	0.12	−0.03
Q6	0.27	0.74	−0.46	0.11	−0.00
Q7	0.77	−0.19	−0.22	−0.24	−0.27
Q8	0.74	−0.15	−0.08	0.19	0.03
Q9	0.35	0.24	0.58	0.37	−0.39
Q10	−0.06	0.12	0.52	0.26	0.26
Q11	0.34	0.28	0.33	0.25	−0.67
Q12	0.79	−0.26	−0.06	0.01	0.14
Q13	0.63	−0.11	0.10	0.53	0.15
Q14	0.81	−0.22	−0.04	−0.18	0.08
Q15	0.76	−0.15	−0.17	−0.24	−0.23
Q16	0.73	−0.02	0.02	−0.02	0.37
Q17	0.61	0.10	0.07	0.41	0.35
Q18	0.12	0.53	−0.49	0.24	0.14
Q19	0.32	0.45	0.39	−0.58	0.20
Q20	−0.03	0.49	0.07	−0.29	−0.14

Recall from Chapter 5 that the principal components can be written as

$$\mathbf{y} = \mathbf{A}'\mathbf{x}, \tag{15.2}$$

where $\mathbf{x}$ is the vector of the original variables, $\mathbf{y}$ is the vector of principal components, and $\mathbf{A}$ is the matrix of the coefficients of the PC's. Now since $\mathbf{A}$ is orthogonal, multiplying equation (15.2) by $\mathbf{A}$ gives $\mathbf{x} = \mathbf{A}\mathbf{y}$.

Now consider the model

$$\mathbf{x} = \mathbf{A}\mathbf{y} + \mathbf{e},$$

where $\mathbf{e}$ is an error vector, and the emphasis of $\mathbf{y}$ changes from that of principal components derived from the original variables to that of underlying latent variables that cannot actually be observed. Also, we suppose that only the first q PC's are used in $\mathbf{y}$ in the model, and correspondingly in $\mathbf{A}$. The principal components are now termed *factors*, and are relabelled $F1, F2, \ldots, F5$. Changing the emphasis from PCs to factors in our example, the questions can now be written in terms of the factors as

$$Q2 = 0.50 \times F1 + 0.12 \times F2 + 0.51 \times F3 + 0.11 \times F4 + 0.02 \times F5 + e_2$$
$$Q3 = 0.78 \times F1 - 0.20 \times F2 - 0.22 \times F3 - 0.29 \times F4 - 0.21 \times F5 + e_3$$

$$\vdots \qquad \vdots \qquad \vdots \qquad \vdots \qquad \vdots \qquad \vdots$$

$$Q20 = -0.03 \times F1 + 0.49 \times F2 + 0.07 \times F3 - 0.29 \times F4 - 0.14 \times F5 + e_{20}.$$

The coefficients in the equations (i.e. matrix $\mathbf{A}$) are called *factor loadings*.

Now write the model in terms of the standardized principal components, $\mathbf{z}$, where

$$\mathbf{z} = \mathbf{\Lambda}^{-1/2}\mathbf{y},$$

where $\mathbf{\Lambda}$ is the diagonal matrix of the first q eigenvalues obtained in the principal components analysis. Then

$$\mathbf{x} = \mathbf{A}\mathbf{\Lambda}^{1/2}\mathbf{z} + \mathbf{e}.$$

Now let $\mathbf{\Gamma} = \mathbf{A}\mathbf{\Lambda}^{1/2}$, and hence the model becomes

$$\mathbf{x} = \mathbf{\Gamma}\mathbf{z} + \mathbf{e}. \tag{15.3}$$

Let the errors be uncorrelated, with $\mathrm{var}(\mathbf{e}) = \mathbf{\Psi} = \mathrm{diag}(\psi_1^2, \ldots, \psi_p^2)$. The variance matrix, $\mathbf{\Sigma}$, of $\mathbf{x}$ is thus given by

$$\mathrm{var}(\mathbf{x}) = \mathbf{\Sigma} = \mathbf{\Gamma}\mathbf{\Gamma}' + \mathbf{\Psi}.$$

Now, if an orthogonal rotation is applied to the factors, to give a new set of factors, $\mathbf{w} = \mathbf{C}\mathbf{z}$ say, where $\mathbf{C}$ is an orthogonal matrix, then the model (15.3) becomes

$$\mathbf{x} = \mathbf{\Gamma}\mathbf{C}'\mathbf{w} + \mathbf{e}$$
$$= \mathbf{\Gamma}^*\mathbf{w} + \mathbf{e},$$

where $\mathbf{\Gamma}^* = \mathbf{\Gamma}\mathbf{C}'$.

The variance matrix of $\mathbf{x}$ is now

$$\mathbf{\Sigma} = \mathbf{\Gamma}\mathbf{C}'\mathbf{I}\mathbf{C}\mathbf{\Gamma}' + \mathbf{\Psi} = \mathbf{\Gamma}\mathbf{\Gamma}' + \mathbf{\Psi},$$

which is exactly the same as before, and thus rotating the factors orthogonally does not affect the variance matrix of $\mathbf{x}$. The factors are not unique. How do we choose an orthogonal rotation? One possibility is to make the factors align more with the original variables, i.e. making a few of the coefficients within the factors as large as possible in magnitude, and the rest small. There have been several suggestions of how this can be achieved. Three are briefly described.

Varimax rotation.   The rotation, $\mathbf{C}$, is found so that $\mathbf{\Gamma}^* = \mathbf{\Gamma}\mathbf{C}'$, and the quantity, V, is maximized, where

$$V = \sum_{j=1}^{q} \left[p^{-1} \sum_{i=1}^{p} (\gamma_{ij}^2/g_i^2)^2 - \left(p^{-1} \sum_{i=1}^{p} (\gamma_{ij}^2/g_i^2) \right)^2 \right],$$

with $[\mathbf{\Gamma}]_{ij} = \gamma_{ij}$ and $g_i = \sum_{j=1}^{q} \gamma_{ij}^2$. The quantity, V, is the sum of the 'variances' of γ_{ij}^2/g_i^2, over j. Since $0 \le \gamma_{ij}/g_i \le 1$, maximizing V tends to make the quantities equal to either zero or unity, and hence the desired result of the rotation.

Quartimax rotation.   The quartimax criterion is similar to the varimax criterion, but the quantity to be maximized is Q, where

$$Q = (pq)^{-1} \sum_{j=1}^{q} \sum_{i=1}^{p} (\gamma_{ij}^2)^2 - \left((pq)^{-1} \sum_{j=1}^{q} \sum_{i=1}^{p} (\gamma_{ij}) \right)^2.$$

Orthomax rotation.   The quantity here to be maximized is

$$O = \sum_{j=1}^{q} \Big[\sum_{i=1}^{p} \gamma_{ij}^4 - \lambda p^{-1} \Big(\sum_{i=1}^{p} \gamma_{ij}^2 \Big)^2 \Big],$$

where $0 \le \lambda \le 1$. For $\lambda = 0$, O becomes the quartimax criterion, Q. When $\lambda = 1$, O becomes the unscaled varimax criterion. For $\lambda = q/2$, the criterion is called the equamax criterion. Note that we are now describing the factor analysis model in terms of the population covariance matrix, Σ. In practice, the sample covariance matrix or correlation matrix is used in place of the population covariance or correlation matrix, and to be consistent with estimation of factor loadings later, we now use $\hat{\Gamma}$ in place of Γ, since they are derived from the data.

Example

Returning to the consumer study data, the principal components were rotated using a varimax rotation. The rotation matrix is

$$\begin{bmatrix} 0.77 & 0.54 & 0.15 & 0.19 & 0.22 \\ -0.31 & -0.07 & 0.76 & 0.50 & 0.26 \\ -0.32 & 0.24 & -0.59 & 0.49 & 0.50 \\ -0.37 & 0.54 & 0.19 & -0.63 & 0.37 \\ -0.28 & 0.59 & 0.01 & 0.26 & -0.71 \end{bmatrix}.$$

Table 15.2 shows $\hat{\Gamma}$.

Table 15.2 The factor loadings after rotation

	F1	F2	F3	F4	F5	Comm
Var	4.18	2.79	2.37	2.13	1.86	13.34
Q2	0.14	0.46	-0.12	0.34	0.42	0.53
Q3	0.90	0.11	0.04	0.07	0.06	0.82
Q4	0.05	0.11	-0.00	0.89	0.10	0.81
Q5	0.10	0.03	0.87	0.09	0.07	0.78
Q6	0.09	0.04	0.90	0.13	0.06	0.85
Q7	0.88	0.09	0.06	0.03	0.12	0.81
Q8	0.56	0.52	0.08	-0.09	0.14	0.62
Q9	-0.02	0.28	-0.04	0.14	0.84	0.80
Q10	-0.42	0.38	-0.17	0.21	0.19	0.43
Q11	0.16	-0.02	0.11	0.04	0.88	0.81
Q12	0.67	0.52	-0.04	0.03	-0.01	0.72
Q13	0.25	0.74	-0.06	-0.18	0.25	0.71
Q14	0.74	0.40	-0.05	0.16	-0.02	0.75
Q15	0.84	0.12	0.06	0.08	0.13	0.74
Q16	0.47	0.61	0.09	0.25	-0.10	0.67
Q17	0.17	0.77	0.21	0.03	0.10	0.68
Q18	-0.04	0.12	0.77	-0.07	-0.09	0.62
Q19	0.14	0.04	0.06	0.90	0.03	0.84
Q20	-0.05	-0.27	0.27	0.42	0.15	0.35

Thus the questions given in terms of the rotated factors are

$$Q2 = 0.14 \times F1 + 0.46 \times F2 - 0.12 \times F3 + 0.34 \times F4 + 0.42 \times F5$$
$$Q3 = 0.90 \times F1 + 0.11 \times F2 + 0.04 \times F3 + 0.07 \times F4 + 0.06 \times F5$$

$$\vdots \quad = \quad \vdots \qquad \vdots \qquad \vdots \qquad \vdots \qquad \vdots$$

$$Q20 = -0.05 \times F1 - 0.27 \times F2 + 0.27 \times F3 + 0.42 \times F4 + 0.15 \times F5.$$

The first factor, $F1$, is associated with questions $\{Q3, Q7, Q12, Q14, Q15\}$, which are to do with drying of the product, and hence we might name the first factor, *'drying'*. The second factor, $F2$, is associated with questions $\{Q13, Q16, Q17\}$, and we might name this *'greasy'*. The main questions for the third factor, $F3$, are $\{Q5, Q6, Q18\}$, and we name this factor *'efficacy'*. The fourth factor, $F4$, is associated with questions $\{Q4, Q19\}$, and we name this factor *'fragrance'*, and lastly, the fifth factor, $F5$, is associated with questions $\{Q9, Q11\}$, and we call this *'application'*. Also given in Table 15.2, are the *communalities*. The communality for variable X_i, is the amount of variance of variable X_i, not explained by the factors. This is measured by

$$\text{var}(X_i) - \psi_i.$$

Typically, the variances for the questions are about 1.0, and so we see that from the estimated communalities, the factors explain less than half of the variation. This is typical for these type of data.

The next step is to predict the factor scores for each of the panellists. There are several ways of doing this. Finding the linear function of $\mathbf{x}$ which minimizes the variance of the prediction error for each factor, gives the predicted scores

$$\hat{\mathbf{z}} = (\hat{\mathbf{\Gamma}}\hat{\mathbf{\psi}}^{-1}\hat{\mathbf{\Gamma}})^{-1}\hat{\mathbf{\Gamma}}'\hat{\mathbf{\psi}}^{-1}(\mathbf{x} - \bar{\mathbf{x}}).$$

Another approach, assuming a multivariate normal distribution, gives predicted scores as

$$\hat{\mathbf{z}} = [\mathbf{I} + (\hat{\mathbf{\Gamma}}\hat{\mathbf{\psi}}^{-1}\hat{\mathbf{\Gamma}})]^{-1}\hat{\mathbf{\Gamma}}'\hat{\mathbf{\psi}}^{-1}(\mathbf{x} - \bar{\mathbf{x}}),$$

which are similar to the previous scores. Let $\mathbf{A} = (\hat{\mathbf{\Gamma}}\hat{\mathbf{\psi}}^{-1}\hat{\mathbf{\Gamma}})^{-1}\hat{\mathbf{\Gamma}}'\hat{\mathbf{\psi}}^{-1}$, or the other expression for calculating the factor scores. Matrix $\mathbf{A}$ contains the factor score coefficients, and allows the matrix of factor scores to be calculated as

$$\mathbf{Z}' = \mathbf{X}\mathbf{A}',$$

where $\mathbf{X}$ is the mean corrected data matrix. Table 15.3 gives the factor score coefficients for the consumer data.

Figure 15.1 shows the factor scores for the second factor plotted against those for the first. The points are labelled according to the product tested by the particular panellist. The three sets of points are intermixed, indicating no differences between the products. This was also true for plots of the scores of the other factors. One point to notice is that where the score on factor 1 is low, the variation is large for the score on factor 2.

As stated previously, there is often confusion between principal components analysis and factor analysis. Principal components analysis models the variances of the random

Table 15.3 Factor score coefficients

	F1	F2	F3	F4	F5
Q2	−0.05	0.14	−0.07	0.12	0.16
Q3	0.27	−0.13	−0.02	−0.00	0.01
Q4	−0.03	0.02	−0.04	0.44	−0.05
Q5	−0.01	−0.02	0.37	0.00	0.02
Q6	−0.02	−0.02	0.38	0.02	0.01
Q7	0.27	−0.15	−0.01	−0.03	0.06
Q8	0.08	0.14	0.01	−0.09	0.02
Q9	−0.06	0.02	−0.03	−0.02	0.46
Q10	−0.21	0.25	−0.07	0.09	0.04
Q11	0.06	−0.17	0.03	−0.08	0.54
Q12	0.12	0.14	−0.04	−0.01	−0.08
Q13	−0.06	0.31	0.02	−0.14	0.06
Q14	0.16	0.06	−0.06	0.05	−0.08
Q15	0.25	−0.12	−0.01	−0.00	0.05
Q16	0.02	0.24	0.01	0.11	−0.18
Q17	−0.10	0.35	0.08	−0.02	−0.06
Q18	−0.06	0.08	0.34	−0.05	−0.07
Q19	0.01	−0.03	−0.01	0.45	−0.08
Q20	0.02	−0.17	0.10	0.19	0.09

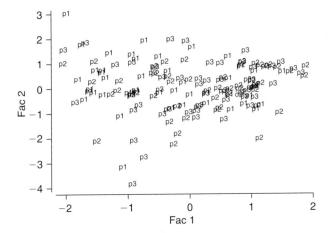

Figure 15.1 Factor scores for the first two factors

variables, so that the first PC explains as much of the total variation in the data as possible, the second PC as much as possible, but uncorrelated to the first, etc. The emphasis is on the diagonal elements of the covariance matrix, $\mathbf{S}$, whereas factor analysis seeks latent variables that underlie the observed variables, and emphasis is on the off-diagonal elements of $\mathbf{S}$.

15.2 The factor analysis model

The factor analysis model, as seen in the previous section, is

$$\mathbf{x} = \boldsymbol{\mu} + \boldsymbol{\Gamma}\mathbf{z} + \mathbf{e},$$

where, this time, we have included the mean vector, μ, of $\mathbf{x}$. We assume $\mathbf{x}$ is a random vector of length p, $\mathbf{z}$ consists of the q factors, and

$$E(\mathbf{z}) = \mathbf{0}, \quad E(\mathbf{e}) = \mathbf{0},$$

$$E(\mathbf{x}) = \boldsymbol{\Sigma}, \quad E(\mathbf{ee}') = \boldsymbol{\Psi}, \quad E(\mathbf{zz}') = \mathbf{I}, \quad E(\mathbf{ze}') = \mathbf{0},$$

with $\boldsymbol{\Psi}$ a diagonal matrix. That is, $\mathbf{z}$ and $\mathbf{e}$ have zero mean vectors, the Z_i's are uncorrelated, each with variance unity, the e_i's are uncorrelated, each with variance Ψ_{ii}, and Z_i and e_j are uncorrelated for all i, j.

As before,

$$\boldsymbol{\Sigma} = \boldsymbol{\Gamma}\boldsymbol{\Gamma}' + \boldsymbol{\Psi}.$$

Now $\boldsymbol{\Sigma}$ has $p(p+1)/2$ variances and covariances, whilst the model has $p(m+1)$ parameters. Thus the model is only useful if $p(m+1) \leq p(p+1)/2$, i.e. if $m \leq (p-1)/2$. As a simple illustrative example for $p = 3$, let

$$\boldsymbol{\Sigma} = \begin{bmatrix} 5 & 2 & 3 \\ 2 & 6 & 6 \\ 3 & 6 & 10 \end{bmatrix}.$$

Then

$$\boldsymbol{\Sigma} = \begin{bmatrix} 5 & 2 & 3 \\ 2 & 6 & 6 \\ 3 & 6 & 10 \end{bmatrix} = \begin{bmatrix} 1 \\ 2 \\ 3 \end{bmatrix} \begin{bmatrix} 1 & 2 & 3 \end{bmatrix} + \begin{bmatrix} 1 & 0 & 0 \\ 0 & 2 & 0 \\ 0 & 0 & 1 \end{bmatrix}.$$

There is only one factor, Z, with unique factor loadings $[1, 2, 3]'$.

Now consider the covariance matrix

$$\boldsymbol{\Sigma} = \begin{bmatrix} 5 & 2 & 3 \\ 2 & 6 & 6 \\ 3 & 6 & 8 \end{bmatrix},$$

very similar to before, but where the variance for X_3 is changed from 10 to 8. This time $\psi_3 = -1$, and although we have found a factor, it is not sensible because of the negative variance. Mathematically there is a solution, but statistically it is not valid. This is known as the *Heywood case*.

In practice, $\boldsymbol{\Sigma}$ will not be known, and so is estimated by the covariance matrix, $\mathbf{S}$. This will lead to estimated factor loadings, $\hat{\boldsymbol{\Gamma}}$. The aim is to find $\hat{\boldsymbol{\Gamma}}$ and $\hat{\boldsymbol{\Psi}}$, so that

$$\hat{\boldsymbol{\Sigma}} = \hat{\boldsymbol{\Gamma}}\hat{\boldsymbol{\Gamma}}' + \hat{\boldsymbol{\Psi}}$$

is satisfied. However, as seen in the previous section, there is (usually) no unique solution, as any orthogonal rotation to $\hat{\boldsymbol{\Gamma}}$ will produce another valid solution.

The solution of the previous section was based on principal components. Here, we assume that $\mathbf{x}$, $\mathbf{z}$ and $\mathbf{e}$ all have multivariate normal distributions. (The assumption of any two of $\mathbf{x}$, $\mathbf{z}$, $\mathbf{e}$, having multivariate normal distributions implies the other must also have a multivariate normal distribution.)

The probability density function for $\mathbf{x}$ is

$$(2\pi)^{-p/2}|\boldsymbol{\Sigma}|^{-1/2}\exp\left\{\tfrac{1}{2}(\mathbf{x} - \mu)'\boldsymbol{\Sigma}^{-1}(\mathbf{x} - \mu)\right\},$$

and hence the log-likelihood is

$$L = \text{constant} - \frac{n}{2}\log|\mathbf{\Gamma\Gamma'} + \mathbf{\Psi}| - \frac{1}{2}\sum_{i=1}^{n}(\mathbf{x}_i - \boldsymbol{\mu})'(\mathbf{\Gamma\Gamma'} + \mathbf{\Psi})^{-1}(\mathbf{x}_i - \boldsymbol{\mu}).$$

The maximum likelihood estimate of $\boldsymbol{\mu}$ is $\bar{\mathbf{x}}$, and substituting this back into the log-likelihood gives, after some algebra,

$$L = -\frac{n}{2}\{\log|\mathbf{\Gamma\Gamma'} + \mathbf{\Psi}| + \text{trace}((\mathbf{\Gamma\Gamma'} + \mathbf{\Psi})^{-1}\mathbf{S})\}.$$

The log-likelihood is minimized iteratively. At each iteration, L is minimized with respect to $\mathbf{\Gamma}$, keeping $\mathbf{\Psi}$ constant, and then minimized with respect to $\mathbf{\Psi}$, keeping $\mathbf{\Gamma}$ constant. If the Heywood case occurs, the offending ψ_i is put equal to a small positive number and the iteration continues. See Mardia et al. (1979) or Krzanowski (2000) for further details of the iterative process. Once the factor loadings have been found, they can be rotated, as before, using one of the methods described above.

The maximum likelihood method was used on the consumer data, together with a varimax rotation. The rotated factor loadings are given in Table 15.4. The first factor, $F1$, is associated with questions $\{Q8, Q12, Q13, Q14, Q16\}$, and to a lesser extent $\{Q2, Q3, Q7, Q15, Q17\}$. We might term the first factor, 'greasy-sticky'. The second factor, $F2$, is associated with questions $\{Q3, Q7, Q15\}$, and we might term this 'drying'. The questions associated with the third factor, $F3$, are $\{Q5, Q6, Q18\}$, and we term this factor 'efficacy'. The fourth factor, $F4$, is associated with questions $\{Q4, Q19\}$, and we term this factor 'fragrance', and the fifth factor, $F5$, is associated with questions $\{Q9, Q11\}$, and we term this 'application'. The last three factors for the maximum likelihood method are very similar to those from the principal components method. The first two factors show some differences in the questions they emphasize. With the maximum likelihood method, it is possible to test hypotheses

Table 15.4 The factor loadings using maximum likelihood

	$F1$	$F2$	$F3$	$F4$	$F5$	Comm
Var	3.72	2.16	2.05	1.83	1.68	11.44
Q2	0.43	0.05	−0.10	0.26	−0.39	0.42
Q3	0.47	−0.79	0.01	0.07	−0.06	0.85
Q4	0.08	−0.03	0.01	0.83	−0.17	0.72
Q5	0.08	−0.09	0.81	0.08	−0.06	0.68
Q6	0.08	−0.08	0.95	0.11	−0.06	0.93
Q7	0.44	−0.75	0.05	0.04	−0.11	0.78
Q8	0.57	−0.34	0.08	−0.04	−0.22	0.49
Q9	0.18	0.09	−0.01	0.11	−0.83	0.75
Q10	0.09	0.37	−0.08	0.06	−0.24	0.21
Q11	0.07	−0.12	0.09	0.08	−0.69	0.50
Q12	0.90	−0.23	−0.01	−0.04	0.01	0.86
Q13	0.56	−0.05	0.02	−0.14	−0.34	0.46
Q14	0.81	−0.34	−0.02	0.13	−0.00	0.80
Q15	0.46	−0.70	0.03	0.10	−0.12	0.72
Q16	0.69	−0.17	0.08	0.15	−0.04	0.54
Q17	0.47	−0.04	0.16	0.03	−0.27	0.33
Q18	0.05	0.01	0.62	0.04	0.05	0.39
Q19	0.15	−0.04	0.05	0.94	−0.03	0.92
Q20	−0.13	−0.01	0.21	0.25	−0.04	0.12

Table 15.5 Values of Λ for various values of q

q	Λ	df	p
0	1653.6	171	<0.0001
1	934.1	152	<0.0001
2	734.6	134	<0.0001
3	524.6	117	<0.0001
4	346.6	101	<0.0001
5	242.3	86	<0.0001
6	121.4	72	0.0002
7	82.9	59	0.0219
8	53.9	47	0.1755

about the number of factors needed in the analysis. If q factors are adequate, then the statistic

$$\Lambda = \{n - (2p + 11)/6 - 2q/3\}\{\log |\hat{\Lambda}\hat{\Lambda}' + \hat{\Psi}| - \log |S|\}$$

has a chi-squared distribution with $\{(p - q)^2 - (p + q)\}/2$ degrees of freedom. Table 15.5 gives the values of Λ for the consumer data for various values of q. This suggests eight factors are required. On the other hand, looking at the variances of the lower order factors, five factors are probably adequate.

As a final analysis, the scores for $Q1$ (overall opinion) were regressed on the scores for the five factors obtained by the principal components method. The regression equation obtained was

$$Q1 = 3.57 + 0.37\,F1 + 0.16\,F2 + 0.34\,F3 + 0.56\,F4 - 0.09\,F5,$$

with R^2 value of 37.2%. The contribution to the regression sum of squares for each of the factors is

$F1$	$F2$	$F3$	$F4$	$F5$
20.1	4.0	16.7	46.1	1.1

Now the factors are uncorrelated, and so the design matrix for the regression is orthogonal. The contribution to the regression sum of squares by each of the factors does not depend upon which factors are included in the model. Thus, we see that the factors $F1$, $F3$ and $F4$ are the only ones required in the model. When the other two factors are removed, the R^2 value reduces to 35.0%. Thus, it appears that *overall opinion* depends, in order, on the factors of *fragrance, drying* and *efficacy*.

15.3 Partial least squares regression

Problems occur in regression analysis (Chapter 10) when there are so many regressor variables (explanatory variables), that multicollinearities occur, and the regression cannot be carried out. There are various ways to overcome this problem. One possibility is to simply choose a small subset of the regressor variables. Another is to perform a principal components analysis on the regressor variables, and then use the most important principal components as the actual regressor variables in the regression. Another approach is to use *partial least squares regression* (PLS) which brings together the ideas of multivariate regression analysis and factor analysis.

For partial least squares regression, the response variables, $Y_1, \ldots, Y_p$ and the regressor variables, $X_1, \ldots, X_q$, are related through two sets of latent variables (or factors), $U_1, \ldots, U_l$ for the Y's, and $T_1, \ldots, T_l$ for the X's. The latent variables are derived linearly, with the U's uncorrelated with each other, the T's uncorrelated with each other, and the correlation of the pairs (U_i, T_i) maximized.

The classic PLS algorithm for finding the latent variables is as follows. First, we assume that all response variables and all regressor variables have been normalized to have zero means and unit variances. As before, let $\mathbf{Y}$ be the matrix of responses, and $\mathbf{X}$, the matrix of regressor values. The pairs, (U_i, T_i), of latent variables are found in turn. Let $T_1 = \mathbf{w}_1' \mathbf{x}$ and $U_1 = \mathbf{c}_1' \mathbf{y}$, where $\mathbf{w}_1$ and $\mathbf{c}_1$ are vectors of weights, normalized such that $\mathbf{w}_1' \mathbf{w}_1 = \mathbf{c}_1' \mathbf{c}_1 = 1$. Let $\mathbf{t}_1$ and $\mathbf{u}_1$ be vectors of scores for the first pair of latent variables. Choose $\mathbf{u}_1$ arbitrarily as a starting vector. Then the following steps are taken

Step 1 $\mathbf{w}_1 = \mathbf{X}' \mathbf{u}_1 / \mathbf{u}_1' \mathbf{u}_1$

Step 2 $\mathbf{w}_1 = \mathbf{w}_1 / \|\mathbf{w}_1\|$

Step 3 $\mathbf{t}_1 = \mathbf{X} \mathbf{w}_1 / \mathbf{w}_1' \mathbf{w}_1$

Step 4 $\mathbf{c}_1 = \mathbf{X}' \mathbf{t}_1 / \mathbf{t}_1' \mathbf{t}_1$

Step 5 $\mathbf{c}_1 = \mathbf{c}_1 / \|\mathbf{c}_1\|$

Step 6 $\mathbf{u}_1 = \mathbf{Y} \mathbf{c}_1 / \mathbf{c}_1' \mathbf{c}_1$

returning to step 1 until convergence.

Upon convergence, the PLS $\mathbf{X}$-loadings are given by

$$\mathbf{p}_1 = \mathbf{X}' \mathbf{t}_1 / \mathbf{t}_1' \mathbf{t}_1,$$

and the residual matrices by

$$\mathbf{E} = \mathbf{X} - \mathbf{t}_1 \mathbf{p}_1' \qquad \mathbf{F} = \mathbf{Y} - \mathbf{t}_1 \mathbf{c}_1'.$$

Once this first pair of latent variables has been found, the second pair (U_2, T_2), is found by repeating the process on the matrices $\mathbf{E}$ and $\mathbf{F}$ in place of $\mathbf{X}$ and $\mathbf{Y}$, respectively. The process continues until l pairs of latent variables are obtained.

The weight vectors, $\mathbf{w}_i$, are placed in a $q \times l$ matrix $\mathbf{W}$, the weight vectors, $\mathbf{c}_i$, are placed in a $p \times l$ matrix $\mathbf{C}$, and the $\mathbf{X}$-loadings vectors are placed in a $q \times l$ matrix $\mathbf{P}$. The PLS regression equation is

$$\mathbf{Y} = \mathbf{X} \mathbf{B} + \mathbf{F},$$

and the PLS estimate of $\mathbf{B}$ is $\mathbf{W}(\mathbf{P}'\mathbf{W})^{-1}\mathbf{C}'$.

Example

The beef and pork consumption data, described in Chapter 10, were subjected to PLS regression using PROC PLS in the statistical software package SAS. The response variables

Table 15.6 X-loadings for the meat consumption data

Factor	pbe	ppo	pfo	dinc	cfo	rdinc	rfp
1	−0.46	−0.21	0.25	0.44	0.48	0.46	0.23
2	0.00	0.63	0.43	0.28	−0.13	0.13	0.56
3	−0.18	0.08	−0.73	−0.38	0.28	0.19	−0.43
4	−0.65	0.12	0.24	−0.21	−0.30	−0.61	0.04

Table 15.7 Weights **W**

Factor	pbe	ppo	pfo	dinc	cfo	rdinc	rfp
1	−0.67	−0.34	0.17	0.34	0.51	0.35	0.12
2	−0.20	1.08	−0.04	0.05	0.05	0.39	0.51
3	−0.31	0.08	−0.72	−0.45	0.22	0.09	−0.39
4	−0.65	0.07	0.21	−0.24	−0.32	−0.60	0.12

Table 15.8 Weights **C**

Factor	cbe	cpo
1	0.80	0.60
2	0.39	−0.92
3	0.72	−0.70
4	0.95	0.30

were *cbe* – consumption of beef, and *cpo* – consumption of pork. The explanatory variables were *pbe* – price of beef, *ppo* – price of pork, *pfo* – retail food price index, *dinc* – disposable income per capita index, *cfo* – food consumption per capita index, *rdinc* – index of real disposable income per capita, and *rfp* – retail food price index adjusted by the CPI.

The number of latent variables (factors) was chosen automatically by SAS, using cross-validation which leaves one observation out of the analysis, in turn, and then predicts its value using PLS regression on the rest of the observations. The accuracy of the resulting predictions determine the number of latent variables to be chosen. Four latent variables were chosen.

The **X**-loadings (i.e. **P**) are given in Table 15.6. The weights, **W**, are given in Table 15.7, and the weights, **C**, in Table 15.8. The estimates of the regression coefficients, **B**, are given by

	cbe	cpo
pbe	−1.10	0.06
ppo	0.69	−1.10
pfo	−0.44	0.62
dinc	−0.36	0.31
cfo	0.23	−0.10
rdinc	0.10	−0.43
rfp	0.13	−0.13

We see that the price of beef, and to a lesser extent the price of pork, is a main driver for the consumption of beef, but for the consumption of pork, the price of beef has very little influence. For this dataset, PLS regression was not really needed since there are no problems with multicollinearity.

15.4 Exercises

1. From the matrix equation

$$\Sigma = \begin{bmatrix} 5 & 1 & 2 \\ 1 & 5 & 2 \\ 2 & 2 & 6 \end{bmatrix} = \begin{bmatrix} \gamma_1 \\ \gamma_2 \\ \gamma_3 \end{bmatrix} \begin{bmatrix} \gamma_1 & \gamma_2 & \gamma_3 \end{bmatrix} + \begin{bmatrix} \psi_1 & 0 & 0 \\ 0 & \psi_2 & 0 \\ 0 & 0 & \psi_3 \end{bmatrix},$$

solve for $\gamma_1, \gamma_2, \gamma_3, \psi_1, \psi_2, \psi_3$. Find the communalities.

2. Repeat Exercise 1 using the covariance matrix

$$\Sigma = \begin{bmatrix} 5 & 1 & 2 \\ 1 & 5 & 2 \\ 2 & 2 & 4 \end{bmatrix}.$$

This is the *Heywood case.*

3. Let covariance matrix Σ be given by

$$\Sigma = \begin{bmatrix} 4 & 1 & 0 & 2 & 4 \\ 1 & 3 & 2 & 2 & 3 \\ 0 & 2 & 10 & 4 & 4 \\ 2 & 2 & 4 & 6 & 6 \\ 4 & 3 & 4 & 6 & 12 \end{bmatrix}.$$

Show that

$$\Gamma = \begin{bmatrix} 1 & 1 \\ 1 & 0 \\ 2 & -2 \\ 2 & 0 \\ 3 & 1 \end{bmatrix}, \quad \Psi = \begin{bmatrix} 2 & 0 & 0 & 0 & 0 \\ 0 & 2 & 0 & 0 & 0 \\ 0 & 0 & 2 & 0 & 0 \\ 0 & 0 & 0 & 2 & 0 \\ 0 & 0 & 0 & 0 & 2 \end{bmatrix}$$

satisfy the factor analysis model

$$\Sigma = \Gamma \Gamma' + \Psi.$$

Now rotate Γ using the rotation matrix

$$C = \begin{bmatrix} 1/\sqrt{2} & 1/\sqrt{2} \\ 1/\sqrt{2} & -1/\sqrt{2} \end{bmatrix},$$

and show that the new factor loadings matrix also satisfies the model.

4. Suppose we scale our data by multiplying variable X_i by b_i, $i = 1,\ldots,p$. Let $\mathbf{B} = \mathrm{diag}(b_i)$, and so the scaling is accomplished as $\mathbf{BX}$. By multiplying the factor analysis model equation,

$$\mathbf{X} = \mu + \mathbf{\Gamma Z} + \mathbf{e},$$

by $\mathbf{B}$, show that $\mathbf{Y} = \mathbf{CX}$ also satisfies the model, with factor loading matrix $\mathbf{B\Gamma}$ and specific variances $b_i^2 \psi_i$. Thus a factor analysis is not affected by the scales of measurements, and in practice, either the sample covariance matrix or the sample correlation matrix can be used for the analysis.

5. Carry out a factor analysis on the US crime rate data. Try the principal components method and the maximum likelihood method. First use two factors and then three. Plot the component scores for one factor against another, labelling the points with the state names.

6. Carry out PLS regression on the following data

x_1	x_2	x_3	x_4	y_1	y_2
20	39	19	59	48	17
24	49	33	73	47	12
32	62	54	94	42	12
12	21	20	33	40	26
44	85	89	129	31	15
40	80	80	120	32	16
34	66	60	100	39	19
22	43	25	65	48	18
19	39	18	58	52	17
45	87	92	132	28	15
34	66	60	100	37	23
16	29	5	45	54	12
14	25	8	39	47	17
31	61	52	92	40	18
20	36	16	56	52	17

Try to carry out ordinary multivariate regression on the same data.

7. Carry out PLS regression on the bodyfat data, described in Chapter 4, using one response variable, percentage bodyfat using Brozek's equation, and fourteen explanatory variables: age, weight, height, adiposity index, and circumferences of the neck, chest, abdomen, hip, thigh, knee, ankle, biceps, forearm, and wrist.

16

Other latent variable models

Chapter 15 introduced *factor analysis* which attempts to find *latent variables* that cannot be observed, that drive the actual *observed variables*, or *manifest variables*. Both the latent and manifest variables were continuous. We now consider the cases where the latent and manifest variables can be categorical. *Latent class analysis* deals with the case where both manifest and latent variables are categorical. *Latent trait analysis* deals with the case where the manifest variables are categorical and the latent variables are continuous. *Latent profile analysis* has continuous manifest variables and categorical latent variables. Only brief details of the models are given here.

16.1 Latent class analysis

We assume that there is only one latent class variable with K classes, with probability function

$$\Pr(Y = k) = \alpha_k \quad \left(k = 1, \ldots, K; \sum_k \alpha_k = 1\right).$$

This can be viewed as a prior distribution on the K classes.

Let X_i have I_i categories. Let π_{ijk} be the probability that category j is observed for variable i, when the latent class is k, i.e.

$$\Pr(X_i = j | Y = k) = \pi_{ijk} \quad \left(\sum_{j=1}^{I_i} \pi_{ijk} = 1\right).$$

Then, for the latent class model formulation, (15.1),

$$f(\mathbf{x}) = \sum_{k=1}^{K} \alpha_k \prod_{i=1}^{p} \prod_{j=1}^{I_i} \pi_{ijk}^{x_{ijk}},$$

where $x_{ijk} = 1$ if the observed category for variable i is j, when the latent class is k, and zero otherwise.

From a sample, $\mathbf{x}_1, \ldots, \mathbf{x}_n$, of observations, the log-likelihood, incorporating Lagrange multipliers to deal with the parameter constraints, is

$$l = \sum_{l=1}^{n} \log\left(f(\mathbf{x}_l)\right) - \lambda \sum_{k=1}^{K} \alpha_k - \sum_{i=1}^{p} \sum_{k=1}^{K} \gamma_{ik} \left(\sum_{j=1}^{l_i} \pi_{ijk} - 1\right), \qquad (16.1)$$

where λ and $\{\gamma_{ik}\}$ are the Lagrange multipliers.

We now differentiate the log-likelihood with respect to α_k and π_{ijk}, to arrive at estimating equations for the maximum likelihood estimates, $\hat{\alpha}_k$ and $\hat{\pi}_{ijk}$ (see Exercises).

The posterior probability that an observation with response $\mathbf{x}$ belongs to latent class k, is

$$p(k|\mathbf{x}) = \alpha_k \prod_{i=1}^{p} \prod_{j=1}^{l_i} \pi_{ijk} / f(\mathbf{x}),$$

which can be estimated using $\hat{\alpha}_k$ and $\hat{\pi}_{ijk}$.

Example

The *Titanic* data described in Chapter 7 was subjected to latent class analysis. The program used to carry out the analysis was CDAS. A useful resource for this general area and for details of software is the website:

http://ourworld.compuserve.com/homepages/jsuebersax/index.htm.

Table 16.1 shows a frequency count of the cells in the four-way contingency table. A latent class variable with three cells was chosen and fitted to the data. The estimated cell conditional probabilities are given in Table 16.2.

Thus the probability of a person being in class 1 of the latent class variable is 0.179, of being in class 2 is 0.565, and for class 3, 0.255. Given a person is in class 1 of the latent variable, the probability of the person being a 1st Class passenger is 0.445, of being adult, 0.899, of being female, 0.846, and being a survivor, 0.972. These are the categories that describe a typical person in class 1 of the latent variable. For those in class 2, the dominant categories are: crew, adult, male, and not surviving. For class 3, the dominant categories are: 3rd Class passenger, adult, male, and not surviving.

From the model, we can calculate expected cell frequencies. For instance, the expected number of female crew members who survived is

$$2201 \times 0.179 \times (0.062 \times 0.899 \times 0.846 \times 0.972)$$
$$+ 0.565 \times (0.686 \times 1.000 \times 0.000 \times 0.200)$$
$$+ 0.255 \times (0.012 \times 0.877 \times 0.242 \times 0.140) = 18.2.$$

Table 16.1 shows the observed, O_i, and expected, E_i cell counts, together with residuals calculated as $(O_i - E_i)/\sqrt{E_i}$. For the most part, the model fits well, the exceptions being the frequencies for: 1st class, female children, surviving; 2nd class, adult males, surviving; 2nd class, male children, surviving. A Pearson chi-squared test would actually reject the model, mainly because of these three badly fitting cells.

Table 16.1 Frequency table for the *Titanic* data

Class	Age	Sex	Survival	Frequency	Expected	Residual
1st	Adult	Female	s	140	130.1	0.87
1st	Adult	Female	ns	4	3.87	0.12
1st	Adult	Male	s	57	53.5	0.48
1st	Adult	Male	ns	118	120.0	−0.18
1st	Child	Female	s	1	14.6	−3.56
1st	Child	Female	ns	0	0.4	−0.65
1st	Child	Male	s	5	2.7	1.44
1st	Child	Male	ns	0	0.1	−0.28
2nd	Adult	Female	s	80	80.2	−0.02
2nd	Adult	Female	ns	13	11.7	0.39
2nd	Adult	Male	s	14	44.5	−4.57
2nd	Adult	Male	ns	154	131.5	1.96
2nd	Child	Female	s	13	9.0	1.32
2nd	Child	Female	ns	0	1.58	−1.26
2nd	Child	Male	s	11	2.3	5.77
2nd	Child	Male	ns	0	4.2	−2.05
3rd	Adult	Female	s	76	80.2	−0.47
3rd	Adult	Female	ns	89	93.7	−0.48
3rd	Adult	Male	s	75	81.6	−0.73
3rd	Adult	Male	ns	387	379.6	0.38
3rd	Child	Female	s	14	9.4	1.49
3rd	Child	Female	ns	17	13.1	1.08
3rd	Child	Male	s	13	7.9	1.80
3rd	Child	Male	ns	35	40.5	−0.86
Crew	Adult	Female	s	20	18.2	0.42
Crew	Adult	Female	ns	3	1.8	0.92
Crew	Adult	Male	s	192	174.4	1.34
Crew	Adult	Male	ns	670	687.3	−0.67

Table 16.2 Estimated cell conditional probabilities

Variable	Category	$\hat{\pi}_{ij1}$	$\hat{\pi}_{ij2}$	$\hat{\pi}_{ij3}$
Class	Crew	0.062	0.686	0.012
	1st	0.445	0.120	0.000
	2nd	0.270	0.102	0.092
	3rd	0.224	0.092	0.896
Age	Adult	0.899	1.000	0.877
	Child	0.101	0.000	0.123
Sex	female	0.846	0.000	0.242
	male	0.154	1.000	0.758
Survival	s	0.972	0.200	0.140
	ns	0.028	0.800	0.860
α_i		0.179	0.565	0.255

16.2 Latent trait analysis

Latent trait analysis is also referred to as *item response theory (IRT)* and the *Rasch model*, although these are particular latent structure models. Here, we introduce the basic Rasch model only.

Suppose that in order to measure mathematical ability of children in a school class, we give them a test consisting of a series of questions, each of which is marked as either correct or incorrect for each child. We assume each child has an individual ability or trait, α_r $(r = 1, \ldots, n)$. The questions have differing levels of difficulty, with the ith question having difficulty parameter β_i $(i = 1, \ldots, p)$.

Let $\mathbf{x}$ be a binary variable, which takes the value unity if a particular question is answered correctly by a child, and zero otherwise. Let the probability that a child answers a question correctly, given 'ability to answer the question', θ, be $\pi(\theta)$. This is called the *item response function (IRF)* or *item characteristic curve (ICC)*. The usual choice for $\pi(\theta)$ is the logistic function, and thus

$$\pi(\theta) = \frac{e^\theta}{1 + e^\theta}.$$

For the rth child, answering the ith question, we model the ability to answer the question as $\theta_{ri} = \alpha_r - \beta_i$. Let $X_{ri} = 1$ if the rth child answers the ith question correctly, and zero otherwise, and hence

$$\Pr(X_{ri} = 1) = 1 - \Pr(X_{ri} = 0) = \frac{e^{\alpha_r - \beta_i}}{1 + e^{\alpha_r - \beta_i}}.$$

We assume answers to questions are independent, and hence the likelihood from all answers by all the children is

$$L = \prod_{r=1}^{n} \prod_{i=1}^{p} \left\{ \frac{e^{\alpha_r - \beta_i}}{1 + e^{\alpha_r - \beta_i}} \right\}^{x_{ri}} \left\{ \frac{1}{1 + e^{\alpha_r - \beta_i}} \right\}^{(1 - x_{ri})}. \tag{16.2}$$

The maximum likelihood estimates $\hat{\alpha}_r, \hat{\beta}_i$ of α_r and β_i can now be found numerically.

Example

Thirty-two children each answered five questions designed to test their knowledge of the kings and queens of England. A small proportion of the data is shown in Table 16.3.

Table 16.3 History test results

Child	Q1	Q2	Q3	Q4	Q5
1	0	1	1	0	0
2	1	0	1	1	0
3	1	0	1	1	0
⋮	⋮	⋮	⋮	⋮	⋮
32	1	0	1	1	0

The Rasch model was fitted using MATLAB. The model is over-parameterized, since adding a constant to each α, and subtracting the same constant from each β, leaves the model unchanged. To overcome this, α_1 was put equal to zero. The maximum likelihood estimates were

Child with no. correct	0	1	2	3	4	5
α_r	−15.43	−1.19	0.00	1.86	3.05	16.06

Question	Q1	Q2	Q3	Q4	Q5
β_i	−0.89	−0.37	0.47	1.44	1.95

Note that for this model, $\hat{\alpha}_r$ depends only on the total number of correctly answered questions, and not on the particular questions. Thus every child that answered every question incorrectly had $\hat{\alpha}_r = -15.43$, those that answered one question correctly had $\hat{\alpha}_r = -1.19$, etc. The β's for the questions ranged from −0.89 for Q1 which was the easiest, to 1.95 for Q5 which was the hardest.

16.3 Latent profile analysis

The manifest variables for latent profile analysis are continuous, and the latent variable is categorical, giving the latent classes. Let there be K latent classes as for the latent class model. Then from (15.1)

$$f(\mathbf{x}) = \sum_{k=1}^{K} \alpha_k \prod_{i=1}^{p} g_i(x_i|k),$$

where $g_i(x_i|k)$ is the conditional probability density of x_i in latent class k.
   The likelihood for a sample is

$$L = \prod_{r=1}^{n} \sum_{k=1}^{K} \alpha_k \prod_{i=1}^{p} g_i(x_{ri}|k),$$

and hence the log-likelihood is

$$l = \sum_{r=1}^{n} \log \left\{ \sum_{k=1}^{K} \alpha_k \prod_{i=1}^{p} g_i(x_{ri}|k) \right\}.$$

For an appropriately chosen distribution, $g(\,\cdot\,)$, the log-likelihood can be maximized.
   The posterior probability that an observation, $\mathbf{x}$, belongs to latent class k, is

$$p(k|\mathbf{x}) = \alpha_k \prod_{i=1}^{p} g_i(x_i|k)/f(\mathbf{x}),$$

which is estimated by replacing the parameters by their maximum likelihood estimates.

Example

Consider the case for $p = 2$, $K = 2$, and $g(\,\cdot\,)$ a normal distribution. Let the mean for x_i, for the kth class be μ_{ik} ($i, k = 1, 2$). Let there be a common variance, σ^2. Then the log-likelihood is given by

$$l = \sum_{r=1}^{n} \log \left[\frac{1}{2\pi\sigma^2} \left\{ \alpha_1 e^{-\frac{1}{2}(x_{r1}-\mu_{11})^2/\sigma^2 - \frac{1}{2}(x_{r2}-\mu_{21})^2/\sigma^2} \right. \right.$$

$$\left. \left. + \alpha_2 e^{-\frac{1}{2}(x_{r1}-\mu_{12})^2/\sigma^2 - \frac{1}{2}(x_{r2}-\mu_{22})^2/\sigma^2} \right\} \right]. \qquad (16.3)$$

The log-likelihood can be maximized using the E-M algorithm, details of which are not given here.

16.4 Exercises

1. Differentiate the log-likelihood for the latent class model of Section 16.1 with respect to α_k and π_{ijk}, and hence derive the likelihood equations.

2. Find the log-likelihood from the likelihood (16.2) for the latent trait model. Differentiate this with respect to α_r and β_i, and hence derive the likelihood equations.

3. Differentiate the log-likelihood (16.3) for the latent profile model of Section 16.3 with respect to μ_{ik} and σ^2, and hence derive the likelihood equations.

17

Graphical modelling

Graphical modelling fits graphs to data. By this, we mean graphs in the sense of graph theory and not graphical illustrations of data. Two books on the subject are Borgelt and Kruse (2002) and Edwards (2000), the latter describing some easy to use software, MIM, which has been used in this chapter for the analyses. However, before explaining graphical modelling, we need to cover some preliminary material regarding *conditional independence* and *graph theory*.

17.1 Conditional independence

Consider two univariate random variables, X_1 and X_2. Recall that the correlation between X_1 and X_2 is given by

$$\rho_{12} = E[(X_1 - \mu_1)(X_2 - \mu_2)]/[\text{sd}(X_1)\text{sd}(X_2)],$$

where $\text{sd}(X_i)$ is the standard deviation of X_i. The sample correlation is

$$r_{12} = \frac{\sum (x_{1i} - \bar{x}_1)(x_{2i} - \bar{x}_2)}{\left[\sum (x_{1i} - \bar{x}_1)^2 \sum (x_{2i} - \bar{x}_2)^2\right]^{1/2}}.$$

For normally distributed random variables, if $\rho_{12} = 0$, then X_1 and X_2 are not only uncorrelated, but also independent.

Now consider three random variables, X_1, X_2, X_3. Figure 17.1 shows a scatterplot matrix for a random sample for these random variables. All three pairs of variables appear to be correlated with each other. We return to this sample later.

The *partial correlation*, $\rho_{12.3}$, of X_1 with X_2, **given** X_3, is

$$\rho_{12.3} = E[(X_1 - \mu_1)(X_2 - \mu_2)|X_3]/[\text{sd}(X_1|X_3)\text{sd}(X_2|X_3)].$$

It can be shown that for the multivariate normal distribution,

$$\rho_{12.3} = \frac{\rho_{12} - \rho_{13}\rho_{23}}{\left[(1 - \rho_{13}^2)(1 - \rho_{23}^2)\right]^{1/2}}.$$

So, if X_1 and X_2, given X_3, are independent, then $\rho_{12.3} = 0$, and hence

$$\rho_{12} = \rho_{13}\,\rho_{23}.$$

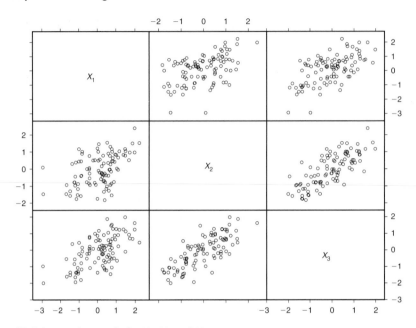

Figure 17.1 Scatterplot matrix for X_1, X_2 and X_3

Now for our sample above, we group the observations according to the values of x_3, using the intervals $(-2.00, -1.10)$, $(-1.10, -0.50)$, $(-0.50, 0.0)$, $(0.0, 0.35)$, $(0.35, 0.90)$, $(0.90, 2.10)$, and draw separate scatterplots of X_1 against X_2 using trellis graphics of S-Plus. These are shown in Figure 17.2. The value of X_3 is approximately constant for each of the scatterplots, and we see that in each case X_1 and X_2 now appear to be uncorrelated, illustrating the notion of conditional independence. (Note that the sample was generated using $\rho_{12} = 0.48$, $\rho_{13} = 0.6$, and $\rho_{23} = 0.8$; i.e. $\rho_{12} = \rho_{13}\rho_{23}$.)

We can extend partial correlations to make the dependence on a subset of variables. Denote this subset by $\mathbf{x_m}$, and then

$$\rho_{12.\mathbf{m}} = \text{corr}(X_1, X_2 | \mathbf{x_m}).$$

Thus for example, $\rho_{12.34...p} = \text{corr}(X_1, X_2 | X_3, X_4 ..., X_p)$.

Let $\mathbf{x} = (X_1, \ldots, X_p)' \sim N_p(\boldsymbol{\mu}, \boldsymbol{\Sigma})$, and

$$\boldsymbol{\Sigma}^{-1} = \boldsymbol{\Omega} = \begin{pmatrix} \omega^{11} & \omega^{12} & \cdots & \omega^{1p} \\ \vdots & \vdots & \ddots & \vdots \\ \omega^{p1} & \omega^{p2} & \cdots & \omega^{pp} \end{pmatrix}.$$

The matrix $\boldsymbol{\Omega}$ is called the *precision matrix*.

It can be shown that the conditional distribution of (X_1, X_2) given $(X_3, \ldots, X_p)$ is bivariate normal with covariance matrix

$$\begin{pmatrix} \omega^{11} & \omega^{12} \\ \omega^{21} & \omega^{22} \end{pmatrix}^{-1} = \frac{1}{\omega^{11}\omega^{22} - (\omega^{12})^2} \begin{pmatrix} \omega^{22} & -\omega^{21} \\ -\omega^{12} & \omega^{11} \end{pmatrix},$$

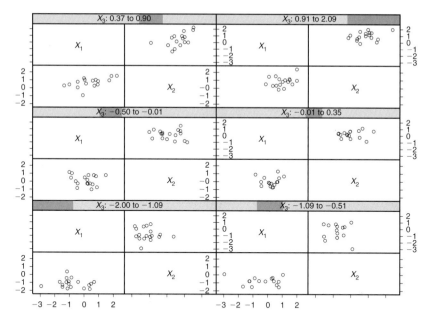

Figure 17.2 Trellis graphics for X_1 and X_2, conditioned on X_3

and hence the partial correlation $\rho_{12.34...p}$ is

$$\rho_{12.34...p} = \frac{-\omega^{12}}{(\omega^{11}\omega^{22})^{1/2}}.$$

Thus two variables are independent, given the remaining variables, if and only if the corresponding element of the inverse covariance matrix is zero. If X and Y are conditionally independent given Z, we write this as

$$X \perp\!\!\!\perp Y|Z.$$

If $X \perp\!\!\!\perp Y|Z$, then the joint probability density function, $f(x, y, z)$, can be factorized as

$$f(x, y, z) = h(x, z)k(y, z),$$

where $h(x, z)$ is a function that involves x and z, but not y, and similarly, $k(y, z)$ involves y and z, but not x. When the conditioning is on more than one variable we write expressions for conditional independence such as $X \perp\!\!\!\perp Y|(W, Z)$, or $X \perp\!\!\!\perp Y|$(the rest of the variables).

17.2 Graphs

A graph, $G = (V, E)$, consists of a finite set, V, of vertices (or nodes), and a finite set, E, of edges (or arcs) between the vertices. Vertices will be labelled X, Y, etc. and for us will represent random variables. Edges will be denoted by $[XY]$, $[YZ]$, etc. For example, Figure 17.3 shows a graph with vertices W, X, Y, Z, and edges $[WX]$, $[XY]$, $[XZ]$, $[YZ]$. Vertices are represented by the symbol o, and edges by a line.

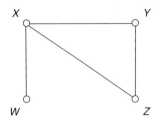

Figure 17.3 A simple graph

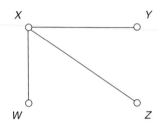

Figure 17.4 A graph showing conditional independence

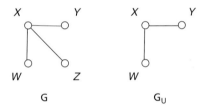

Figure 17.5 An example of a subgraph

In fact, it is common practice for continuous variables to be represented by the symbol o, and for discrete variables by the symbol •.

Two vertices are *adjacent* if there is an edge between them, i.e. if $[XY] \in E$, and we will write $X \sim Y$. So, in the graph of Figure 17.3, $W \sim X$, but $W \not\sim Z$. A graph is *complete* if there is an edge between every pair of vertices. We have said that vertices will represent random variables. but what will edges represent? This leads us on to *independence graphs*.

Independence graphs. An independence graph has vertices representing random variables, and edges which are only present for pairs of variables which are **not** conditionally independent, given the rest of the variables. Thus we have for all pairs (X, Y), where $X \perp\!\!\!\perp Y|$(the rest), the edge between vertices X and Y omitted, otherwise it is retained. For example, the graph in Figure 17.4 has $W \perp\!\!\!\perp Z|(X, Y)$, $Y \perp\!\!\!\perp Z|(X, W)$ and $W \perp\!\!\!\perp Y|(X, Z)$.

Consider a graph $G = (V, E)$. Any subset $U \subset V$ induces a *subgraph* of G, $G_U = (U, F)$, where F consists of those edges in E where both endpoints are in U. Figure 17.5 shows an example where $G = (\{W, X, Y, Z\}, \{[WX], [XY], [XZ],\})$ and $G_U = (\{W, X, Y\}, \{[WX], [XY]\})$.

A subset $U \subset V$ is called *complete* if it induces a complete subgraph. Figure 17.6 shows two subgraphs from the graph G shown in Figure 17.5. The subgraph with vertices $\{W, X\}$ is complete, while the subgraph with vertices $\{W, X, Y\}$ is not.

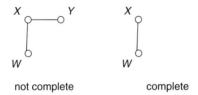

Figure 17.6 Two subgraphs, only one of which is complete

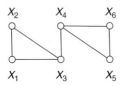

Figure 17.7 A graph with three cliques

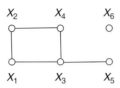

Figure 17.8 An unconnected graph with X_1, X_2, X_3, X_4 a chordless 4-cycle

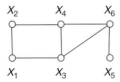

Figure 17.9 $\{X_3, X_4\}$ separates $\{X_1, X_2\}$ and $\{X_5, X_6\}$

A subset $U \subset V$ is called a *clique* if it is *maximally complete*. That is, if U is complete, then for any other set, W, containing U, W is not complete. The cliques in the graph shown in Figure 17.7 are $\{X_1, X_2, X_3\}$, $\{X_4, X_5, X_6\}$, and $\{X_3, X_4\}$. So, for instance, $\{X_1, X_2\}$ cannot be a clique, because $\{X_1, X_2, X_3\}$ is complete. A graph can be represented by, and constructed from, its cliques.

A sequence of vertices, X_0, X_1, $\ldots$, X_n, such that $X_{i-1} \sim X_i$ is called a *path* between X_0 and X_n, of length n. (Note that here, the labelling of the vertices as X_0 to X_n is for convenience in order to denote the sequential path from one vertex to another.) A graph is *connected* if there is a path between every pair of vertices.

A path $X_1, X_2, \ldots, X_n, X_1$ is called an *n-cycle*. If the n vertices of an *n*-cycle, $X_1, X_2, \ldots, X_n$, are distinct, and if $X_j \sim X_k$ only if $|j - k| = 1$ or $n - 1$, then we call it a *chordless n-cycle*. The graph in Figure 17.8 is not connected since there is not a path between X_6 and any of the other vertices. The sequence of vertices, X_1, X_2, X_4, X_3, forms a chordless 4-cycle.

For three subsets U, W and S of V, we say S *separates* U and W, if all paths from U to W intersect S. For example, in Figure 17.9, let $U = \{X_1, X_2\}$, $W = \{X_5, X_6\}$, and $S = \{X_3, X_4\}$. Then S separates U and W.

17.2.1 Independence models

Recall that in a graph with random variables represented by the vertices, if the edge between vertices X and Y is omitted, then this represents the fact that $X \perp\!\!\!\perp Y|$(the rest). This is the *pairwise Markov property*. In Figure 17.3, $Y \perp\!\!\!\perp W|$(the rest), and $Z \perp\!\!\!\perp W|$(the rest).

Can we reduce the set of variables '(the rest)' down to a subset? It can be shown that if two sets of variables, U and W, are separated by a third set S, then $U \perp\!\!\!\perp W|S$. This is the *global Markov property*. Thus in Figure 17.3, $\{X\}$ separates $\{Y\}$ and $\{W\}$, and so $Y \perp\!\!\!\perp W|X$. Similarly, $\{X, Y\}$ separates $\{W\}$ and $\{Z\}$, and so $W \perp\!\!\!\perp Z|\{X, Y\}$. It can be shown that the pairwise and global Markov properties are equivalent under some general conditions.

A graph can be constructed from its cliques, and in the software package MIM, graphs are defined by expressions such as

$$//WX, XYZ$$

$$//WX, XY, XZ$$

which would give rise to the graphs in Figures 17.3 and 17.4.

17.3 Fitting graph models to data

We assume that the data are from a multivariate normal distribution. The likelihood is

$$L = \prod_{i=1}^{n} (2\pi)^{-p/2} |\boldsymbol{\Sigma}|^{-1/2} \exp\left\{ -\tfrac{1}{2}(\mathbf{x}_i - \boldsymbol{\mu})'\boldsymbol{\Sigma}^{-1}(\mathbf{x}_i - \boldsymbol{\mu}) \right\}.$$

For the *full* or *saturated model*, where all edges are present in the graph and hence no conditional independencies, then $\boldsymbol{\Sigma}$ contains no constraints, and has maximum likelihood estimate

$$\hat{\boldsymbol{\Sigma}}_s = \frac{n-1}{n} \mathbf{S}.$$

Example

Figure 17.10 shows the graph for the saturated model for the blood pressure data described in Chapter 3.

The maximum likelihood estimate of the $\boldsymbol{\Sigma}$ is

$$\hat{\boldsymbol{\Sigma}}_s = \begin{pmatrix} 394.7 & 133.6 & 346.1 & 223.4 \\ 133.6 & 102.4 & 142.9 & 157.5 \\ 346.1 & 142.9 & 373.5 & 302.8 \\ 223.4 & 157.5 & 302.8 & 795.8 \end{pmatrix}.$$

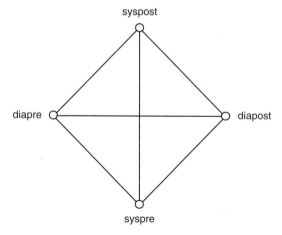

Figure 17.10 Graph of the saturated model

If we make some of the variables conditionally independent, then this implies that corresponding elements, w^{ij}, in $\Omega = \Sigma^{-1}$ are set to zero. Then in maximizing the likelihood, we impose the constraint that $\hat{w}^{ij} = 0$, for each pair of variables, X_i, X_j, that are conditionally independent. Suppose the maximum likelihood for Σ is now $\hat{\Sigma}_m$, under the model. Details of the constrained maximization are not given here.

A measure of whether the model fits or not is given by the deviance, D, where

$$D = 2\big(\log L(\hat{\Sigma}_s) - \log L(\hat{\Sigma}_m)\big).$$

The deviance, D, has an approximate chi-squared distribution with degrees of freedom equal to the number of edges taken away from the saturated model, to arrive at the actual model.

For the blood pressure data, let the vertices be: $a = syspre$, $b = diapre$, $c = syspost$, and $d = diapost$. Edges were removed from the saturated model, and the subsequent model fitted, in a stepwise manner, giving the following deviances.

Edge removed	deviance	df
ab	0.15	1
bd	1.18	2
ad	2.43	3
cd	7.97	4
bc	19.42	5
ac	44.52	6

We choose the model with edges ab, bd, and ad removed. The model is denoted by $//ac$, bc, cd and is shown in Figure 17.11. The maximum likelihood estimate of Σ under the model is

$$\hat{\Sigma}_m = \begin{pmatrix} 394.7 & 132.4 & 346.1 & 280.6 \\ 132.4 & 102.4 & 142.9 & 115.8 \\ 346.1 & 142.9 & 373.5 & 302.8 \\ 280.6 & 115.8 & 302.8 & 795.8 \end{pmatrix},$$

which leads to a deviance of 2.43, on three degrees of freedom ($p = 0.4872$).

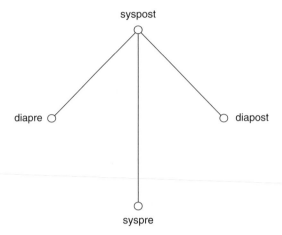

Figure 17.11 Graph of *//ac, bc, cd*

From Figure 17.11, the interpretation of the graphical model is that syspost is not independent of syspre, diapre and diapost, but diapost is conditionally independent of syspre and diapre, given syspost.

17.4 Categorical data

From Chapter 14, we saw how loglinear models can be used to model the probabilities associated with the cells of contingency tables. For example, for a three-way table, the probabilities for the saturated model are

$$np_{ijk} = e^{\theta+\alpha_i+\beta_j+\gamma_k+\alpha\beta_{ij}+\alpha\gamma_{ik}+\beta\gamma_{jk}+\alpha\beta\gamma_{ijk}}.$$

Denote the three categorical variables for the contingency table as A, B, and C (corresponding to α, β, and γ, respectively). For B to be conditionally independent of C, given A, i.e. $B \perp\!\!\!\perp C|A$, we simply put the two-factor interactions, and associated higher interactions, equal to zero in the model. Thus

$$np_{ijk} = e^{\theta+\alpha_i+\beta_j+\gamma_k+\alpha\beta_{ij}+\alpha\gamma_{ik}},$$

and hence

$$np_{ijk} = e^{\theta+\alpha_i+\beta_j+\alpha\beta_{ij}}e^{\gamma_k+\alpha\gamma_{ik}}.$$

Essentially, the probability is now a product of two factors, implying that $B \perp\!\!\!\perp C|A$. See Edwards (2000) for further details, and for the case of mixed models where some variables are continuous and some categorical.

Example

The *Titanic* survival data (Chapter 7) were analysed using graphical modelling. The variables are *Class, Sex, Age, Survived*. Figure 17.12 shows the graph for the saturated model.

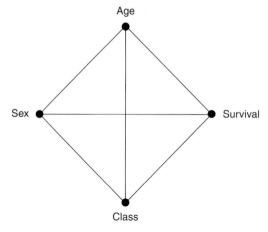

Figure 17.12 Graph for Titanic data

No edges can be removed to give a graphical model that fits the data. Our conclusion is that survival depends on Class, Sex and Age.

Example

The DASL website has a dataset (popular kids) of the responses of children to the question 'What would you most like to do at school?', with possible answers, 'make good grades', 'be good at sports', and 'be popular'. This variable we call *goals (e)*. The children were also asked to rank 'make good grades', 'being good at sports', 'being handsome or pretty', and 'having lots of money', in priority. For our purposes, we use the aspect which they rank as being most important, as a variable which will be named *priority (f)*. We have chosen a subset of the observations and the variables recorded, namely *gender (a)*, *age (b)*, *race (c)*, and *area (d)* where their school is situated. After analysis using MIM, the model chosen was //*cd*, *bef*, *dbe*, *af*, shown in Figure 17.13. Children's goals are conditionally independent of gender, given their priorities, and conditionally independent of race given area and age. Children's priorities are conditionally independent of race and area, given their goals and age.

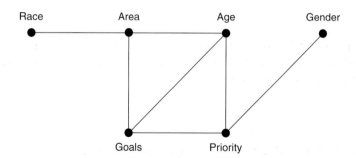

Figure 17.13 Graph for children's goals and priorities

17.5 Exercises

1. Draw the graphs of the following models

$$//XY, XYZ \qquad //W, X, Y, Z \qquad //ab, bcd, cde$$

2. Fit graphical models to Fisher's iris data (Chapter 13).
   Fit graphical models to the soil data of Chapter 6, using the variables *silt, clay, organic matter*, and *pH*.

18

Data mining

Suppose you were the managing director of a chain of supermarket stores, and that you were a stickler for old-fashioned ways. You do not believe in using computers because humans can give a more friendly, reliable and personal service. You employ hundreds of staff. These staff ring prices through the tills, price every item on sale, restock the shelves, stock-take the shelves, and order goods from the warehouse. Staff at the warehouse process orders, stock-take, order goods from suppliers, and so on.

One day, a large commercial concern approaches you with the offer of a contract for an exclusive deal, where your chain of supermarkets will supply their luxury items, bringing substantial profit to both parties. The problem is that they want you to give them a breakdown of number of customers per week, the age structure of your customers, a profile of the customers that buy goods from the luxury end of the market, the typical amount that is spent per week by a customer, an analysis of the periodicity of sales over the months of the year, a forecast of the trend in the number of customers over the next five years, and so on. You and a rival chain of supermarkets have two weeks to supply this information, and then a decision will be made as to which chain will get the contract. Before the two weeks are up, you resign. You know that the rival supermarket chain has up-to-date computers and huge databases from which they can easily supply the information required. You do not even know where to start finding the information.

In our modern world, computer power and ability to store data is now breathtaking compared to just a few decades ago. Huge databases exist now and more are being created all the time. Computers can process data very fast. However, the collection of data and its storage is pointless, unless you can put it to some use. Data are different from information. Data are the collection of facts and figures, while information is the meaningful data extracted from them. Knowledge concerns the interrelations and patterns within the data.

Consider what is happening at the rival supermarket chain. As the goods that a customer is buying are swiped through the till, a record of everything that is being purchased is being stored in a database. If the customer has a store card, then these new purchases are stored alongside all the other purchases that have been made by this customer in the past. The computer is keeping a profile for this customer which it continually updates. From this profile, special offers and personalized marketing material can be sent to the customer. If the customer purchases a lot of wine, why not send them details of the discounts available next month on wine and spirits? If the customer never buys items for babies, do not send them details of special offers on baby goods.

As the items are swiped through the till, stock records are automatically updated. When a certain level of stock is reached for an item, the computer automatically places an order for more from the warehouse. The financial side of buying and selling goods, wages, profit, budgeting, and forecasting can all be done using the data stored on the computer. Various analyses can be carried out on the databases. Associations between items can be measured, where two items are highly associated if the purchase of one of them implies that there is a high probability that the other will be purchased. For example, cigarettes and cigarette lighters are probably highly associated. The classic quote is that nappies (diapers) and beer are highly associated, since the father is sent to purchase bulk quantities of nappies, but while at the supermarket, decides to re-stock with beer.

Banks and other financial institutions store huge databases. All financial transactions can be recorded and stored. As soon as a person opens an account, data on them starts to be collected. A profile is built up. Can you obtain credit? It depends upon your credit score which is based on your financial profile. Industrial processes are often controlled to a large extent by computers. Sensors record temperature, pressure, flow rates, pH, viscosity, gas levels, etc. Hundreds of process measurements might be made every second. These can all be stored. Every ten years a census is carried out in the UK. By law, every household has to return a completed questionnaire relating to the persons living there. The questions relate to the size of the dwelling, employment, etc. This is a large database. There are large medical and biological databases, for instance the mapping of genomes for various species. The list is endless.

18.1 KDD

The acronym KDD stands for *knowledge discovery in databases*, and covers the whole process of extracting knowledge from data. It covers the areas of database technology, statistics and machine learning, expert systems, and data visualization. Data mining is the process of extracting knowledge from the data. By some, data mining is thought of as 'just doing statistics, but on large sets of data'. This is not true. Although 'statistics' plays a major part, data mining and KDD include other aspects. Storing data in the correct manner, making it available to many users, automatically updating, ensuring portability between systems, is not a trivial task, and is just as important as the statistical aspects. Also, it is not possible or nonsensical to carry out certain statistical procedures on huge datasets. For instance, for a dataset with one million observations on one thousand variables, it would not be sensible to attempt to derive a one million by one million dissimilarity matrix. Similarly, to carry out a two-sample t-test with sample size equal to one million would be a pointless exercise, since the test would always find a minute difference between population means as being significant, and the only relevancy would be whether the difference in means was of practical significance.

KDD can be broken down into various parts. *Data selection*: first, it has to be decided which data are to be stored. Usually, not all the data from processes that generate thousands of values per second will need to be stored. Perhaps just the mean value per minute or hour. Whereas, for a bank account, every transaction will need to be stored. For how long? Ten years? Twenty years? One hundred years? Data need to be *cleaned*. Errors need to be detected wherever possible and corrected. Much of this can be done automatically, and due to the scale of the database, might be the only way of dealing with errors. For instance, automatic checks on the ranges of variables can be made, or spelling for categorical data.

Checks on relationships between variables can be made, for instance date of birth and age. Duplication of records can be a problem. For example, a misspelling of a name can cause a problem, and so it appears that AB Grimshaw and AB Grimeshaw both live at a certain address. Data need to be *coded*, so that it can be stored sensibly in the database. Numerical data are usually easily stored. However, decisions might be needed as to how the data are to be stored. For instance, is income to be stored as a number, 25,000 for instance, or is it to be stored as 'middle income'? Perhaps both should be stored, or perhaps a subsequent data mining program will convert 25,000 to 'middle income'. How will addresses be stored? The whole address, or as 39_AB12_3CD, which is a house number followed by a postcode? How do you cope with houses that have a name but no street number?

A database can be for *operational purposes*, for instance the database of prices and stock levels for the goods in our supermarket chain. As each item is swiped through the till, the price is retrieved from the database, the stock level is updated, and the customer profile is updated. Or the database can be for *strategic decision making*. For instance, the database that builds our customer profiles and contains historic sales data can be used to help decide future advertising strategy.

Data warehousing is an important part of the process. Will the database be stored on a remote server and clients download it each time they need to use it? Will the processing be carried out remotely also? Who will update the database? How do you protect the database from malicious damage?

Once a database has been established, it can be put to its purpose. If it is a strategic database, it can be data mined. The simplest way to do this is to 'ask questions' of the database. A common way of doing this is by using SQL (Structured Query Language), which will extract from the database answers to questions. For instance, when asked, it will produce a table of customers, categorized by the local areas, by the mean amount they spend per week. Only answers to the questions posed will be given. However, data mining goes further. It attempts to extract knowledge from the database, without specific questions being asked. It has to be given some direction of course. It could be used to segment (cluster) the customers according to the products they purchase. It could be used to find the major associations between products.

Having mined the database, the next important steps are *visualization of data* and *reporting*. As with any analysis, results are best illustrated with a graph if possible, and a succinct way of reporting the results.

Here we cannot delve any further into the realms of database technology and the other 'non-statistical' aspects of KDD. We can only give a flavour of the statistical side. There are several books on KDD and data mining. Adriaans and Zantinge (1996) give a good non-mathematical introduction; Hand et al. (2001) is for those who wish for a more statistical approach.

18.2 Data mining

Data mining covers all techniques that extract knowledge from large datasets. There tends to be a different slant on the techniques from those who approach data mining from the computer science arena, than those from the statistics arena. For instance, 'supervised learning' is not (normally) a term used by a statistician, who would instead say 'discriminant analysis'. For 'unsupervised learning', the statistician would say 'cluster analysis'. Commonly used techniques that are used in data mining are *regression analysis, principal components analysis, classification and regression trees (CART), discriminant analysis and classification, neural networks*, and *support vector machines (SVM)*. Of course any appropriate statistical technique can be utilized. Many of these techniques have been covered already; this chapter gives an introduction to some more of them.

With many of the models used for data mining, there is the danger of over-fitting (as with all statistical models). It is important to use one subset of the data with which to fit the model (the training data), one to validate the model and tune the model (the validation data), and one to test the final model (test data). If this is not done, it is often possible to find a model that will fit the training data exactly, but is useless when used on other data. In this situation, the model is over-fitted. A model needs to fit the training data well, but also be good at prediction for other data. With huge databases, sampling of the observations has to be carried out in order to produce the training, validation and test data, and it is often impractical to use all the observations because of the enormity of the database.

18.2.1 Rules and patterns

Rules or *models* for databases are summaries or descriptions of the data. For instance, a cluster analysis on consumer data may segment or partition the customer base into various types of customer, ranging from 'most valued customers', to 'occasional purchasers'. *Patterns* within a database are local features, where the data appear to behave differently from the norm. For instance, has the weekly on-line shopping order from a particular customer suddenly changed dramatically, suggesting a change in the household circumstances, such as a birth? Or have the withdrawals from a bank account suddenly changed in character, suggesting a bank card theft?

18.2.2 Neural networks

Neural networks were meant to be a mathematical modelling technique that act like the neurons and synapses of the brain. In reality, they are simply a means of non-linear regression and discrimination/classification. Nevertheless, they are an interesting tool to work with. Bishop (1995) and Ripley (1996) are two books on the subject. Here, we can only give a brief introduction concentrating upon the *multilayer perceptron*.

Figure 18.1 shows a multilayer perceptron. The circles represent the neurons of the brain, but we will call them *nodes*. The synapses are the connections between the neurons; we call them *edges* or *connections*. The first layer of nodes of the perceptron is the *input layer*, where data $x_1, \ldots, x_n$ are entered. Note, for this perceptron, there are five input nodes and so our data will be five-dimensional. The input layer of nodes then outputs to the second layer of nodes known as the *hidden layer*, since the outputs of this layer are not observed directly. The hidden layer of nodes outputs to the *output layer*. This layer is essentially the response variable, **y**. In our perceptron, the response variable is two-dimensional.

Next, we need to model how a node in one layer outputs to a node in the next layer. In the brain, neurons fire or do not fire, and thus output is binary, say 0 or 1. For the perceptron, we allow a smoother output, modelled as a sigmoid function, and called the *activation function*,

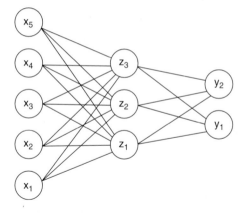

Figure 18.1 Multilayer perceptron

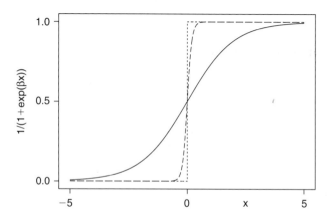

Figure 18.2 Activation functions

$g(x)$. A common function used for this is the logistic function. Thus, the output from a node that has input x, is

$$\frac{1}{1 + e^{-\beta x}}.$$

Figure 18.2 shows three activation functions. The one represented by the solid line is the logistic function with $\beta = 1.0$. The long-dashed line is for the logistic with $\beta = 10.0$, and the short-dashed line is the *Heaviside function*, which has a step change from 0 to 1 at the origin. It can be seen that the larger the value of β, the closer the activation function is to the Heaviside function.

Each of the connections between the input nodes and the hidden layer nodes is given a weight. The weight between the ith input node and the jth hidden layer node is denoted by w_{ij}^I. The total input into the jth hidden layer node is the weighted sum of the outputs from the input nodes,

$$\sum_i w_{ij}^I x_{ij}.$$

The output from the jth node in the hidden layer is

$$z_j = \frac{1}{1 + \exp\{-(w_{0j}^I + \sum_i w_{ij}^I x_{ij})\}},$$

where a *bias* term, w_{0j}^I, has been introduced in order to remove the symmetry within the nodes. The bias can be thought of as an extra node in the input layer whose value is always unity, and has connection weight w_{0j}^I.

The nodes in the hidden layer then feed forward to the output nodes in a similar manner, although a different activation function can be used. The weights for the connections between the hidden layer nodes and the output nodes are w_{jk}^H. The output at node k is

$$y_k = \frac{1}{1 + \exp\{-(w_{0k}^H + \sum_j w_{jk}^H z_{jk})\}}$$

if the same activation function is used as for the nodes in the hidden layer.

18.2.3 Training the network

Once the neural network has been designed with the number of input, hidden, and output nodes chosen, it has to be *trained*, i.e. the weights w_{ij}^I, w_{jk}^H have to be estimated from training data. We want the network to give outputs that match target values, which we will denote by Y_i. If there is only one output node (i.e. only one response variable), then the weights can be estimated by those values of w_{ij}^I, w_{jk}^H, that minimize

$$E = \sum_i (y_i - Y_i)^2,$$

the total squared error between the output and the target values. For multiple output nodes, this can be generalized to

$$E = \sum_i \sum_k (y_{ik} - Y_{ik})^2,$$

where y_{ik} is the output for the ith training sample at the kth node, and Y_{ik} is the target value for the ith training sample for the kth variable. To put it in a statistical context, a multiple regression problem would have k input nodes, one for each of the explanatory variables, X_i, and one output node, the response variable, Y. For a discrimination problem, there would be p input nodes, one for each X_i, and g output nodes, with target values as group indicators.

Various algorithms exist for estimating the weights, by minimizing E, or other error measuring function; see Bishop (1995) for further details.

Example

As an illustration a neural net was used on a small dataset on air pollution and mortality from the DASL website. The data came from the US Department of Labor Statistics with measurements made on factors that could affect mortality for sixty US cities. Age adjusted mortality was the response variable, and there were sixteen explanatory variables. Only five of these variables were selected for use in the neural net. These were: mean January temperature (*JaT*), annual rainfall (*Rain*), median education (*Ed*), percentage of non-whites (*NWhite*), and sulfur dioxide pollution potential (*S02Pot*).

The statistical software package JMP was used for the neural net modelling, using three hidden layer nodes. The hidden layer nodes had the logistic function for the activation function, while the output nodes had the identity function. JMP uses the optimization function, *E*, but also has the option of an over-fit penalty, which tries to restrain the neural net from over-fitting the model. The input variables are scaled to have zero mean and standard deviation unity. JMP will fit the neural net several times in order to try to overcome the problem of local minima. The sixty observations were split into a training set of forty-eight, and a validation set of twelve.

Table 18.1 gives the estimates of the parameters (weights) for the hidden nodes and the output node.

The value of *E* was 4.33, and the penalty function had value 0.43. Current packages do not produce standard errors for parameters, and the parameter estimates are not clearly interpretable. For a neural network, it is more important to assess its performance by using validation and test data. Looking at the nodes separately, the main variable contributing to Hidden 1 is Rain, and to a lesser extent, Ed. To Hidden 2, it is NWhite, and to a lesser extent JaT and S02Pot, and to Hidden 3 Rain and to a lesser extent NWhite and S02Pot. For the output node Hidden 2 contributes more than Hidden 1 and Hidden 3. It must be said that many models gave a similar fit with very different parameter estimates.

Figure 18.3 shows a plot of actual mortality against that predicted from the model. The crosses represent observations in the training set, and the circles, observations in the validation set. As expected, the mortality for the observations in the training set is better predicted than that for the points in the validation set. Figure 18.4 shows a similar plot, but

Table 18.1 Parameter estimates for the neural net

Node	Parameter	Estimate
Hidden 1	intercept	−1.76
Hidden 1	JaT	−1.32
Hidden 1	Rain	5.69
Hidden 1	Ed	2.71
Hidden 1	NWhite	−0.98
Hidden 1	S02Pot	−2.34
Hidden 2	intercept	−0.27
Hidden 2	JaT	0.34
Hidden 2	Rain	−0.16
Hidden 2	Ed	−0.11
Hidden 2	NWhite	−0.70
Hidden 2	S02Pot	−0.24
Hidden 3	intercept	2.02
Hidden 3	JaT	1.92
Hidden 3	Rain	−14.89
Hidden 3	Ed	−2.85
Hidden 3	NWhite	5.81
Hidden 3	S02Pot	4.52
Output	intercept	5.28
Output	Hidden 1	−2.37
Output	Hidden 2	−6.51
Output	Hidden 3	−2.48

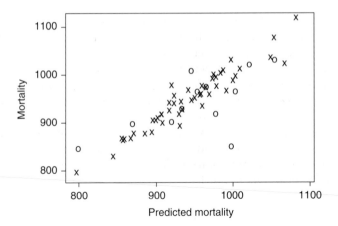

Figure 18.3 Mortality versus predicted mortality

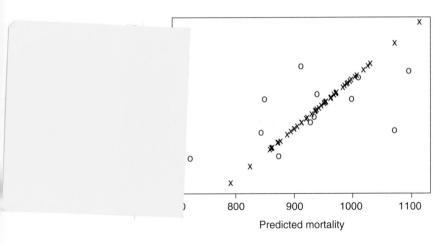

Figure 18.4 Mortality versus predicted mortality showing over-fitting

for a neural network where all sixteen explanatory variables have been used as inputs to the neural network. This time E was 0.00, and the penalty function 0.06. The over-fitting is very apparent!

18.2.4 Projection pursuit

Projection pursuit is a method that seeks *interesting* projections of multivariate data onto a low-dimensional space, usually of one or two dimensions. What is meant by an interesting projection? A projection that shows the data as univariate normal for a projection to a one-dimensional space (an axis), or as bivariate normal for a two-dimensional projection, is considered *uninteresting*. Interesting projections are those that deviate from normality as much as possible.

First, the data are *sphered*, where not only is each variable transformed to have zero mean and variance unity, but correlations are forced to have the value zero. This can be achieved by transforming the mean centered data matrix, to $(\frac{n-1}{n}\mathbf{S})^{-1/2}\mathbf{X}$, where $\mathbf{S}$ is the sample covariance matrix. Now, let $\mathbf{a}$ be a particular projection onto an axis, given by $z = \mathbf{a}'\mathbf{x}$, with $\mathbf{a}'\mathbf{a} = 1$. The projections of the observations are $z_i = \mathbf{a}'\mathbf{x}_i$, $i = 1, \ldots, n$.

A measure of how interesting is the projection, is given by a *projection pursuit index*. There are several of these to choose from, see for example Jones and Sibson (1987). One index is the *entropy index*, $-\int f \log f$, where f is the density function of z. The minimum of the index occurs for the standard normal distribution. By maximizing the index, we try to get as far away as possible from normality. In practice, f is estimated by a kernel density estimator (see Silverman, 1986), $\hat{f}$, such as

$$\hat{f}(z) = n^{-1} \sum_{i=1}^{n} \phi\{(z - z_i)/\lambda\},$$

where ϕ is the standard normal density function, and λ is a smoothing parameter.

The ideas extend to two-dimensional, or higher, projections.

18.3 Text analysis

Analysing textual data is much harder than analysing numerical data. The relationships between words, written or spoken, is very complex. Not only does each word refer to a person, object or a non-tangible entity, or to an action, etc., but the relationship of words to each other conveys a deeper level of communication. *Natural Language Processing (NLP)* is the science of analysing language. For a statistical approach to NLP, see the books by Manning and Schütze (1999) and Oakes (1998).

A *corpus* (plural: *corpora*) is a body of text. Many corpora are now stored in electronic form, which gives us the power to automatically analyse them. Some examples of electronically stored corpora are novels and plays, scientific papers, e-mail messages, Parliament proceedings, and a host more. To analyse the text within the corpus electronically, the corpus has first to be *tokenized*, where every word and punctuation mark is separated out, and is called a *token*. Searching for white space and new line characters will do this, although phrases like 'blood test' can cause problems, since these two words are really a single entity. The words can also be *tagged* as to which part of speech they are. A *lexicon* or electronic dictionary can be used to match words with the same meaning or that are different parts of speech from the same word.

Within a corpus, we wish to know what is being written about. Obviously, you could read the text and answer such a question. But, suppose you wanted to find scientific papers that mentioned wind energy. In order to save time you would use a search engine that can scan through thousands of papers, generally looking at titles and key words. One problem is that wind has two meanings, one the movement of air in the atmosphere, and the other to coil, as in winding up a watch.

Collocations are groups of words that occur together in the same context. For instance 'running scared' and 'ice cream'.

A *concordance* for a word (or phrase) lists the words surrounding the word at every occurrence of the word. For instance for *PCA or PC* from Chapter 5, the concordance, or *KWIC, keywords in context*, is

... aim of principal components analysis	(PCA)	. In general, let the variables ...
... called the first principal component	(PC)	, Y_2 is the second principal ...
... to them as principal components (or	PC's)	.
... Example	PCA	of body fat data
... One problem with	PCA	is that it is not scale ...
etc.	etc.	etc.

The KWIC can be used to understand the contexts in which the keyword is being used. For example the keyword 'cancer' might be used in a medical context or an astrological context.

We can apply various statistical techniques to words (tokens) in a corpus. A starting point is to form a frequency table for the number of times words are used. An early suggested numerical relationship between words in a corpus is *Zipf's law*, which states that if all the words in the frequency table are placed in descending order of frequency, then

$$freq \propto \frac{1}{rank}.$$

If we can have a measure of dissimilarity between words, then multidimensional scaling and clustering techniques can be used to group words. An example of a dissimilarity measure is the number of characters that separate two particular words, averaged over the all the co-occurrences of them within one thousand characters, say.

We can list the words and frequency of use within a corpus, but that does not necessarily tell us what messages are actually being conveyed. To go further, grammar has to be taken into account with the structure of relationships between nouns, pronouns, verbs, adverbs, adjectives, conjunctions, prepositions, articles, etc. One goal is to model context free grammars, where the pattern between the parts of speech is modelled probabilistically.

N-grams are nth order Markov chains, used to model the 'next' word in a sentence. Let the words to be used in the text be denoted by $w_1, w_2, \ldots, w_C$. We think of a process that steps through the words of the sentence. Let the present word used in the sentence be denoted by w_t, the word before that, w_{t-1}, the word before that, w_{t-2}, etc. The next word to be used is denoted by w_{t+1}. The probability that the next word to be used is modelled as

$$p_{i,s} = \Pr(w_{t+1} = w_i | w_t, w_{t-1}, \ldots, w_{t-n+1}),$$

where $s = \{w_t, w_{t-1}, \ldots, w_{t-n+1}\}$. Thus the probability that the next word is w_i depends only on the previously used n words up to the present point. For instance we might assign some probabilities as follows:

$$\Pr(\text{mat} | \text{cat, sat, on, the}) = 0.9$$

$$\Pr(\text{mat} | \text{rain, in, Spain, falls}) = 0.0$$

$$\Pr(\text{mat} | \text{need, to, vacuum, the}) = 0.1.$$

The collection of these probabilities, known as *transition probabilities*, gives rise a Markov chain which can be used model the text. In practice we would not try to model all the words in a corpus, but use certain keywords. The sequences of words in the corpus can be used to estimate the transition probabilities.

Example

The text of twenty Shakespeare plays was downloaded from the website http://www-tech.mit.edu/Shakespeare/ in html format. The file for each play was cleaned of superfluous html commands and punctuation marks using a *Perl* program. (See http://www.perl.org for information about Perl.) Another Perl program was used to produce a frequency count for all the words used for each play. Then 166 of the most frequently used words were chosen that could be used to characterize the plays. Plurals and different parts of speech for the same word were amalgamated into one word. Each play would typically use over two thousand words and so the very commonly used words such as *the, and, I, in*, etc., and proper names were excluded. Table 18.2 gives the keywords used. Table 18.3 shows a small portion of the frequency count data. The full dataset are to be placed on the DASL website.

The plays used were:

Asy –	As You Like It	*Lov* –	Love's Labours Lost
Mer –	Merchant of Venice	*Tam* –	Taming of the Shrew
Tem –	The Tempest	*Twe* –	Twelfth Night
Cym –	Cymbeline	*H41* –	Henry IV, part I
H42 –	Henry IV, part II	*H5* –	Henry V
H8 –	Henry VIII	*R2* –	Richard II
R3 –	Richard III	*Ham* –	Hamlet
Mac –	Macbeth	*Ort* –	Othello
RJ –	Romeo and Juliet	*Tit* –	Titus Andronicus
JC –	Julius Caesar	*Tim* –	Timon of Athens

Table 18.2 Keywords used in Shakespeare plays

age	air	alone	archbishop	banish	battle
beauty	bed	blood	blunt	body	boy
brother	captain	castle	cave	child	citizen
clown	comfort	conscience	court	cousin	crown
daughter	day	dead	deed	devil	die
drink	drown	duke	duty	ear	earl
earth	enemies	eye	face	fair	faith
fall	father	field	fight	fire	fly
fool	fortune	foul	free	friend	gentle
gentleman	ghost	god	gold	grave	grief
ground	hand	hang	happy	hate	head
heart	heaven	help	home	honest	honour
hope	horse	hostess	house	husband	justice
kill	kind	king	kingdom	kiss	knave
knight	lady	land	law	lay	lie
life	lord	love	mad	march	marriage
meet	men	mercy	merry	messenger	money
moor	mother	murder	music	news	night
noble	nurse	old	pardon	peace	pistol
pity	plague	poet	poor	power	pretty
prince	princess	proud	queen	roman	senator
servant	shame	sister	slaughter	sleep	soldier
son	sorrow	soul	sovereign	spirit	sun
swear	sweet	sword	tears	tender	tent
thought	tongue	tonight	truth	uncle	valiant
villain	virtue	war	weep	wife	wish
wit	woe	woman	woo	work	world
worth	wrong	young	youth		

Table 18.3 Frequency count of Shakespeare keywords (only part of the frequency table)

Word	Asy	Lov	Mer	Tam	Tem	Twe	Cym	H41	H42	H5	H8	R2	R3	Ham	Mac	Oth	RJ	Tit	JC	Tim
age	10	1	2	4	4	2	4	6	5	4	2	9	13	9	2	1	6	10	6	4
air	1	6	3	1	12	5	13	3	3	9	3	5	6	12	14	4	7	2	4	8
alone	6	2	3	7	2	10	8	9	5	4	7	3	3	10	3	4	16	4	6	4
archbishop	0	0	0	0	0	0	0	9	37	2	11	1	6	0	0	0	0	0	0	0
banish	12	0	0	1	2	1	11	10	3	0	2	23	1	0	1	1	21	4	3	7
battle	1	0	0	1	0	0	6	7	2	14	0	2	13	0	2	2	0	0	8	1
beauty	3	15	2	9	2	6	4	3	0	2	7	1	9	6	0	5	13	1	1	0
bed	4	0	6	13	4	11	17	5	7	5	5	4	8	11	11	22	22	4	7	2
blood	11	14	16	3	5	17	21	30	31	43	7	46	67	26	38	23	24	30	34	11
blunt	0	1	1	3	0	0	0	26	6	1	0	1	11	0	1	0	0	0	2	0
body	6	3	7	4	2	0	13	4	13	8	4	7	10	17	3	6	11	12	15	1
boy	14	18	18	11	4	12	31	11	23	33	9	8	17	4	1	1	11	31	12	1
brother	45	0	1	1	18	16	30	18	22	25	6	13	63	18	1	4	7	50	13	6
captain	0	0	0	0	0	16	9	2	17	32	0	5	4	9	0	3	1	3	0	6
castle	0	0	0	0	0	0	0	4	2	1	0	16	5	13	16	10	0	1	0	0
cave	3	0	0	0	0	0	15	0	0	0	0	0	1	0	0	0	2	4	0	13
child	6	11	5	6	5	0	8	4	4	6	9	10	39	5	12	3	19	20	4	2
citizen	1	0	1	2	0	0	1	0	0	3	1	0	38	0	0	1	7	2	78	0
clown	5	3	0	0	0	115	0	0	0	0	0	0	0	47	0	17	0	14	0	0
comfort	5	3	0	2	9	3	11	1	4	3	9	11	14	1	5	5	10	4	2	4
conscience	0	1	10	0	2	2	9	1	1	13	22	3	13	8	0	3	0	1	1	3
court	17	15	15	5	2	2	25	8	16	2	18	10	1	6	5	3	2	12	1	1
cousin	12	0	1	1	0	7	0	30	23	14	0	40	19	4	6	2	19	2	0	0
crown	5	4	1	7	8	3	5	9	15	29	7	23	21	7	9	2	1	2	15	5

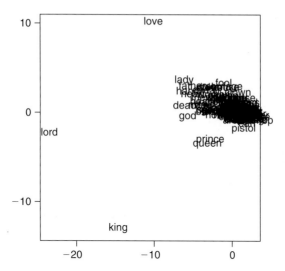

Figure 18.5 Classical scaling of keywords

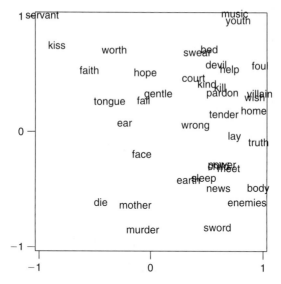

Figure 18.6 A subset of the keywords

First, the keywords were considered as points in a 20-dimensional space, one dimension for each play. The frequency counts for the keywords were used to calculate dissimilarities between all pairs of words using Euclidean distance, and then these were subjected to classical scaling (Chapter 9). Figure 18.5 shows the two-dimensional configuration from the scaling. It is difficult to plot the configuration in such a small page area, and the only thing that can be discerned is that the words *love*, *lord*, and *king* are well removed from the others. These were omitted from the analysis, and the classical scaling repeated. Figure 18.6 shows a subset of the words that were close to the origin.

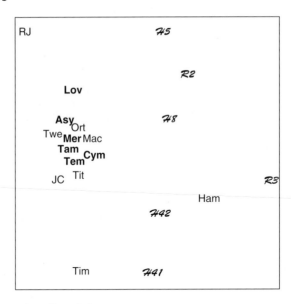

Figure 18.7 Non-metric scaling of plays

Next, dissimilarities were found between all pairs of plays, again using Euclidean distance, but viewing the twenty plays as points within a 166-dimensional space. This time nonmetric scaling, in two dimensions, was used to find a configuration of points representing the plays. The STRESS was 10%. Figure 18.7 shows the configuration of plays, where bold font represents Shakespeare's comedies, italic font represents his historical plays, and normal font, his tragedies.

18.4 Exercises

1. How could web 'chat sites' for young people be policed automatically?

2. A company has an e-mail address where customers ask questions and make comments about products. Typical questions might be, 'Where can I purchase your new range?', 'How much does a super-saw cost? Is there a discount for 10?', 'Your drill burnt out. Can I have a replacement?', and 'How do I make sure my holes are square?' How could an automatic e-mail reply service be set up?

References

Adriaans, P. and Zantinge, D. (1996). *Data Mining*. Harlow: Addison-Wesley.

Agresti, A. (2002). *Categorical Data Analysis*, 2nd Edition. New York: Wiley.

Andrews, D. F. and Herzberg, A. M. (1985). *Data*. New York: Springer-Verlag.

Bartholomew, D. and Knott, M. (1999). *Latent Variable Models and Factor Analysis*, 2nd Edition. London: Edward Arnold.

Bartholomew, D. J. (1987). *Latent Variable Models and Factor Analysis*. Oxford: Griffin.

Basford, K. E. and Tukey, J. W. (1988). *Graphical Analysis of Multi-Response Data: Illustrated with a Plant Breeding Trial*. Boca Raton: Chapman & Hall/CRC.

Bishop, C. M. (1995). *Neural Networks for Pattern Recognition*. Oxford: Clarendon Press.

Borg, I. and Groenen, P. J. F. (1997). *Modern Multidimensional Scaling: Theory and Applications*. New York: Springer-Verlag.

Borgelt, C. and Kruse, R. (2002). *Graphical Models: Methods for Data Analysis and Mining*. Chichester: Wiley.

Breiman, L., Friedman, J. H., Olshen, R. A. and Stone, C. J. (1984). *Classification and Regression Trees*. Belmont: Wadsworth.

Burges, C. J. C. (1998). A tutorial on support vector machines for pattern recognition. *Data Mining and Knowledge Discovery*, 2:955–74 (http://citeseer.ist.psu.edu/burges98tutorial.html).

Carroll, J. D. and Chang, J. J. (1970). Analysis of individual differences in multidimensional scaling via an n-way generalization of "Eckart-Young" decomposition. *Psychometrika*, 35:283–319.

Christianini, N. and Shawe-Taylor, J. (2000). *An Introduction to Support Vector Machines*. Cambridge: Cambridge University Press.

Collett, D. (2003). *Modelling Binary Data*, 2nd Edition. Boca Raton: Chapman & Hall/CRC.

Cox, T. F. and Cox, M. A. A. (2000). *Multidimensional Scaling*, 2nd Edition. Boca Raton: Chapman & Hall/CRC.

Davenport, M. and Studdert-Kennedy, G. (1972). The statistical analysis of judgements: an exploration. *J. R. Stat. Soc., C*, 21:324–33.

Dempster, A. P., Laird, N. M. and Rubin, D. B. (1977). Maximum likelihood from incomplete data via the EM algorithm (with discussion). *J. R. Stat. Soc., B*, 39:1–38.

Dobson, A. J. (2001). *An Introduction to Generalized Linear Models*, 2nd Edition. Boca Raton: Chapman & Hall/CRC.

Edwards, D. (2000). *Introduction to Graphical Modelling*, 2nd Edition. New York: Springer-Verlag.

Everitt, B., Landau, S. and Leese, M. (2001). *Cluster Analysis*, 4th Edition. London: Edward Arnold.

Gabriel, K. R. (1971). The biplot graphical display of matrices with application to principal components model. *Biometrika*, 58:453–67.

Gordon, A. D. (1999). *Classification*, 2nd Edition. Boca Raton: Chapman & Hall/CRC.

Gower, J. C. (1971). A general coefficient of similarity and some of its properties. *Biometrics*, 27:857–74.

Gower, J. C. and Hand, D. J. (1996). *Biplots*. Boca Raton: Chapman & Hall/CRC.

Greenacre, M. J. (1993). *Correspondence Analysis in Practice*. London: Academic Press.

Hand, D. J., Daly, F., McConway, K., Lunn, D. and Ostrowski, E. (1994). *A Handbook of Small Data Sets*. London: Chapman & Hall/CRC.

Hand, D. J., Mannila, and Smyth, P. (2001). *Principles of Data Mining*. Cambridge, Mass.: MIT Press.

Hartingan, J. A. (1975). *Clustering Algorithms*. New York: Wiley.

Hastie, T., Tibshirami, R. and Friedman, J. (2001). *The Elements of Statistical Learning*. New York: Springer-Verlag.

Hsu, J. C. (1996). *Multiple Comparisons: Theory and Methods*. London: Chapman and Hall/CRC.

Innes, J. R. M., Ulland, B. M., Valerio, M. G., Petrucelli, L., Fishbein, L., Hart, E. R., Pallotta, A. J., Bates, R. R., Falk, H. L., Gart, J. J., Klein, M., Motchell, I. and Peters, J. (1969).

Bioassay of pesticides and industrial chemicals for tumorigenicity in mice. *J. Natl. Cancer Inst.*, 42:1101–14.

Jackson, J. E. (1991). *A User's Guide to Principal Components*. New York: Wiley.

Johnson, N. L. and Kotz, S. (1969). *Distributions in Statistics: Discrete Distributions*. New York: Wiley.

Johnson, N. L. and Kotz, S. (1972). *Distributions in Statistics: Continuous Multivariate Distributions*. New York: Wiley.

Jolliffe, I. T. (2002). *Principal Components Analysis*. New York: Springer-Verlag.

Jones, M. C. and Sibson, R. (1987). What is projection pursuit? *J. R. Statist. Soc. A*, 150:1–36.

Kendall, M. (1975). *Multivariate Analysis*. London: Griffin.

Kohonen, T. (1982). Self-organized formation of topologically correct feature maps. *Biological Cybernetics*, 43:59–69.

Kohonen, T. (1990). The self-organizing map. *Proceedings of the IEEE*, 78:1464–80.

Krzanowski, W. J. (2000). *Principles of Multivariate Analysis: A User's Perspective*. Oxford: Oxford University Press.

Krzanowski, W. J. and Marriott, F. H. C. (1994). *Multivariate Analysis, Part 1: Distributions, Ordination and Inference*. London: Edward Arnold.

Krzanowski, W. J. and Marriott, F. H. C. (1995). *Multivariate Analysis, Part 2: Classification, Covariance Structures and Repeated Measurements*. London: Edward Arnold.

Kutner, M. H., Nachtsheim, C. J. and Neter, J. (2003). *Applied Linear Regression Analysis*, 4th Edition. London: McGraw-Hill/Irwin.

Lance, G. N. and Williams, W. T. (1967). A general theory of classificatory sorting strategies: 1. Hierarchical systems. *Comp. J.*, 9:373–80.

Manning, C. D. and Schütze, H. (1999). *Foundations of Statistical Natural Language Processing*. Cambridge, Mass.: MIT Press.

Mardia K. V., Kent, J. T. and Bibby, J. M. (1979). *Multivariate Analysis*. London: Academic Press.

Marriott, F. H. C. (1971). Practical problems in a method of cluster analysis. *Biometrics*, 27:501–14.

McCullagh, P. and Nelder, J. A. (1989). *Generalized Linear Models*, 2nd Edition. London: Chapman & Hall/CRC.

McLachlan, G. J. (1992). *Discriminant Analysis and Statistical Pattern Recognition*. New York: Wiley.

Montgomery, D. C. and Peck, E. A. (1992). *Introduction to Linear Regression Analysis*. New York: Wiley.

Oakes, M. P. (1998). *Statistics for Corpus Linguistics*. Edinburgh: Edinburgh University Press.

Ripley, B. D. (1996). *Pattern Recognition and Neural Networks*. Cambridge: Cambridge University Press.

Roberts, G., Martyn, A. L., Dobson, A. J. and McCarthy, W. H. (1981). Tumour thickness and histological type in malignant melanoma in New South Wales, Australia. 1970–76. *Pathology*, 27:763–70.

Rousseauw, J., du Plessis, J., Benade, A., Jordaan, P., Kotze, J., Jooste, P. and Ferreira, J. (1983). Coronary risk factor screening in three rural communities. *South African Medical Journal*, 64:430–6.

Sammon, J. W. (1969). A nonlinear mapping for data structure analysis. *IEEE Trans. Comput.*, 18:401–9.

Sibson, R. (1978). Studies in the robustness of multidimensional scaling: Procrustes statistics. *J. R. Stats. Soc., B*, 40:234–8.

Silverman, B. (1986). *Density Estimation for Statistics and Data Analysis*. London: Chapman & Hall/CRC.

Stamey, T., Kabalin, J., Johnstone, J., Freiha, F., Redwine, E. and Yang, N. (1989). Prostate specific antigen in the diagnosis and treatment of adenocarcinoma of the prostate ii. Radical prostatectomy treated patients. *Journal of Urology*, 16:1076–83.

Weisberg, S. (1985). *Applied Linear Regression*, 2nd Edition. New York: Wiley.

Willerman, L., Schultz, R., Rutledge, J. N. and Bigler, E. (1991). In vivo brain size and intelligence. *Intelligence* 15: 223–8.

Index